The Fractal Cosmos

The Gossamer Threads of the Cosmos

Stephen Blaha Ph. D.
Blaha Research

MMXXV

.

ISBN: 979-8-9986071-4-1

Rev. 00/00/01 August 23, 2025

To Margaret

Some Other Books by Stephen Blaha

SuperCivilizations: Civilizations as Superorganisms (McMann-Fisher Publishing, Auburn, NH, 2010)

Unification of God Theory and Unified SuperStandard Model THIRD EDITION (Pingree Hill Publishing, Auburn, NH, 2018).

The Exact QED Calculation of the Fine Structure Constant Implies ALL 4D Universes have the Same Physics/Life Prospects (Pingree Hill Publishing, Auburn, NH, 2019).

Passing Through Nature to Eternity ProtoCosmos, HyperCosmos, Unified SuperStandard Theory (Pingree Hill Publishing, Auburn, NH, 2022).

HyperCosmos Fractionation and Fundamental Reference Frame Based Unification: Particle Inner Space Basis of Parton and Dual Resonance Models (Pingree Hill Publishing, Auburn, NH, 2022).

The Cosmic Panorama: ProtoCosmos, HyperCosmos, Unified SuperStandard Theory (UST) Derivation (Pingree Hill Publishing, Auburn, NH, 2022).

God and and Cosmos Theory (Pingree Hill Publishing, Auburn, NH, 2023).

Newton's Apple is Now The Fermion (Pingree Hill Publishing, Auburn, NH, 2023).

Cosmos Theory: The Sub-Particle Gambol Model (Pingree Hill Publishing, Auburn, NH, 2023).

Cosmos-Universe-Particle-Gambol Theory (Pingree Hill Publishing, Auburn, NH, 2024).

Structure and Dynamics of Cosmos Theory and the Unified SuperStandard Theory (Pingree Hill Publishing, Auburn, NH, 2024).

Black Holes, White Holes, and Superluminal Starship (Pingree Hill Publishing, Auburn, NH, 2025).

The Sequence Theory Foundation of Cosmos Theory and Gambol Theory (Pingree Hill Publishing, Auburn, NH, 2025).

BEYOND TIME AND SPACE Fermion Strata & Consistency Conditions for Sequences, Cosmos Spaces and Gambols (Pingree Hill Publishing, Auburn, NH, 2025).

The Cosmos as a Physics Organism (Blaha Research, Auburn, NH, 2025)

The Root of Reality (Blaha Research, Auburn, NH, 2025)

Available on Amazon.com, bn.com, Amazon.co.uk and other international web sites as well as at better bookstores.

CONTENTS

FIGURES and TABLES

Introduction

This book extends the understanding of Cosmos Theory, as presented previously by the author, to a new deeper fractal formulation. Particles and interactions are now composed of fractals embodying growth from the infinitesimal to their form in Cosmos Theory.

The author has previously suggested that Physics is a simulation – a Physical description of Reality based on the simple observation that a change in Physics theory does not cause a change in Reality. They differ except that Physics attempts to describe Reality. In that it mimics Reality, it is a simulation.

This book shows that Physics as currently formulated, including Cosmos Theory as presented previously by the author, has a deeper, very different, formulation. This formulation effectively reduces all the fundamental components of Physics to fractal curves. This new view is a major step in understanding Physical phenomena. The parts of Physical theories had been viewed as "things" that exist in "bulk": particles with size and mass; interactions that are either present or not, and so on. The new fractal view is to treat components as composed of many fractional parts (fractals) that combine to give the components that appear in Physical theories. This reduction to fractals has the benefit of establishing continuity between a new level of infinitesimals and the components of current Physical Theories (including Cosmos Theory before now.)

When one talks of the origin of components, it is more comfortable to see a gradual origin from infinitesimals rather than a magic origin through "popping" from nowhere. This view applies to current theory as well as to the origin of everything.

The origin of the Cosmos appears to be understandable as a fractal generated from a zero dimension point – essentially nothing. We find the set of possible spaces is specified in the construction of a fractal curve from line segments. The Cosmos Fractal Curve has two strands – somewhat analogous to DNA.

This book provides a new fractal paradigm for the inner nature of Physics theories. It is based on infinitesimals that provide a graceful infinitesimal "beginning" to a very large Cosmos. The "beginning" is a conceptual theorization of the Reality of a Cosmos that is without beginning or end.

The new fractal formulation of Cosmos Theory is a new deeper simulation of the previous simulation, which was based on fundamental fermions, interactions, and boson fields. The previous simulation was ultimately based on The Standard Model and its extensions. The depth of the new fractal simulation solidifies the view that Physics is a simulation – composed of a fractal simulation within the conventional Physical simulation.

We will see that the fractal form of Cosmos Theory gives results previously found and adds a new deeper level of features. Some new results are: fractal arithmetic combinations used in Physics applications, a more detailed form of the Cosmos fractal curve that is double stranded, and the transformation of fermions and interactions into a fractal form.

1. A Fractal View of the Cosmos

Past fundamental Physics theories, including Cosmos Theory until recently, have been based on "concrete" Physical entities: fundamental fermions, fundamental gauge fields, interactions, Higgs particles, and so on.

In this book we will consider a new view of these "concrete" entities in which they are composed of infinitesimal parts that combine to form fractal "curves." There are a number of reasons why this new formulation is important:

1. The Cosmos fractal curve is the basis for the generation of the Cosmos Theory spaces.
2. The construction of the Cosmos curve specifies each piecewise linear curve such that each linear curve segment, itself, corresponds to a fractal.
3. The basis of the Master Equation, from which all things arise, is fractal.
4. The dimension arrays of Cosmos spaces may be viewed as fractal.
5. The entries in the Cosmos table of sequences support a fractal interpretation.
6. The fermions, gambols and strata support a fractal interpretation. They each have a fractal substructure.
7. Interactions support a fractal interpretation.

For these reasons we now proceed to map each Cosmos Theory construct to a fractal form. We begin with a brief look at fractals and then proceed to consider the Physical "things" that, until now, were viewed as distinct entities:

Entities and their fractal equivalents
Double Stranded Fractal Cosmos Curve
Fractal view of the Master Equation
Fractal fundamental fermion fields
Fractal fundamental gauge fields
Fractal Interactions and Coupling Constants
Fractal view of Mass
Fractal view of Universes

We will see that the fractal form of Cosmos Theory gives the results previously found and adds a new deeper set of features.

2. Double Stranded Fractal Cosmos Curve

2.1 Brief View of Fractals

The dimension D of simple and fractal curves is called the Hausdorff dimension. It can be an integer, as it is for simple spaces such as 3-dimension space where D = 3. It can be a non-integer value for a space filling curve such as the Hilbert curve where the line occupies a two dimension space with N = 2.

2.1.1 Hausdorff Dimension

The Hausdorff dimension D of a curve is

$$D = \log(N)/\log(\varepsilon) = \log_{\varepsilon}(N) \tag{2.0}$$

where the numerator = log N equals the logarithm of the number of copies of the shape formed from each copy of the previous iteration of the curve under construction. The denominator = log ε is the logarithm of the factor by which the copies are scaled downward from the previous iteration.

For example, a Sierpinski triangle (double sided) creates 3 copies resulting in N = 3 copies from the previous iteration. The scale is ε = 2 based on the doubling of the side. As a result:

$$D = \log(3)/\log(2) = \log_2(3) = 1.585$$

or

$$2^D = 3$$

For the Cosmos Curve and the Hilbert curve we find N = 4 and ε = 2 with the result:

$$D = \log(4)/\log(2) = 2$$

2.1.2 Fractal Cosmos Spaces and Fractal Dimension Arithmetic

The Cosmos Fractal Curve is described (created) in a series of steps – *conceptually.* In Reality it exists instantaneously. There is no point in time for its creation since time does not exist "before" the Cosmos Fractal Curve. So the "creation" process that we describe has the sole purpose of providing an understanding of the Cosmos. The Cosmos Fractal Curve fills a two dimension unit square. It has fractal dimension 2.

At each step of the Cosmos Fractal Curve construction there is a *numerical* value 2^n – the length of the piecewise linear curve used in construction. The number, denoted n, has only been used as the linear curve length until now. We will now expand its definition based on its value being related to the number of components (dimensions) in a Cosmos space. A HyperCosmos space of dimension r has a dimension array with

2^{r+4} components where n = r + 4. A 2nd Kind HyperCosmos space has a dimension array with 2^{r+3} components with n = r + 3.

We will extend the nature of the piecewise linear curve number n to have an associated Cosmos space with a dimension array size of $2^n = 2^{r+4}$. The definition of fractal dimension (including topological dimensions) may be applied to Cosmos spaces by setting N in eq. 2.0 with

$$N = 2^n$$

and ε = 2 with the result that

$$D(n) = \log(2^n)/\log(2) = \log_2(2^n) = n \qquad (2.0a)$$

is the Fractal (topological) dimension of the associated space.

Each Cosmos space now has an associated fractal of fractal dimension $D_C(r) = r + 4$ (HyperCosmos spaces) or $D_{C2}(r) = r + 3$ (2nd Kind HyperCosmos spaces). We are justified in associating construction step n with the Cosmos space of dimension n since there is a 1:1 map between construction steps and Cosmos spaces.

In addition to associating a Cosmos space with each construction step we will also associate a space-time[1] of dimension r with the construction step. These space-times do not exist until universes are created. They become the space-times of the universes. As an associate of a construction step a space-time is an artifact of Cosmos structure and thus not Physically instantiated. See Fig. 2.6 for a diagram of a section of the Cosmos Fractal Curve together with its associated spaces.

At each construction step there is the length of the piecewise linear curve, a Cosmos space and a space-time.

We will find it necessary to "add", "subtract", "multiply", and "divide" fractals. When we perform arithmetic operations they will be applied to fractal dimensions although they will be exemplified with Cosmos dimension arrays. We define fractal dimensions from Cosmos dimension array sizes. Since the fractal dimensions are based on logarithms as in eq. 2.1 below the arithmetic operations on fractal dimensions will be based on logarithms of dimension array sizes. Thus, for example, fractal dimension addition is based on products of dimension array sizes. See eq. 2.3 below.

Fractal dimension arithmetic (in these discussions) is based on dimension array sizes and the nature of logarithms, which lead to a different form for arithmetic operations.

2.1.2.1 Fractal Addition

The addition of fractals does not seem to have been considered previously. Later we will see fractal addition is implied in fractal relations between particles. We will study fractal addition in the case of Cosmos spaces.

There are several definitions of dimensions for Cosmos spaces:

1. We have assigned an even integer dimension r to each Cosmos space.

---

[11] For Cosmos space of dimension r there is a subset of r dimensions for its space-time in its dimension array.

2. Each HyperCosmos space of dimension r has a dimension $D(2^{r+4})$ equal to the number of elements in its dimension array $d_{dr} = 2^{r+4}$.
3. Each 2nd Kind HyperCosmos space of dimension r has a dimension $D(2^{r+3})$ equal to the number of elements in its dimension array $d_{dr} = 2^{r+3}$.
4. Each HyperCosmos space and 2nd Kind HyperCosmos space of dimension r has a space-time dimension $D(2^r) = r$.

We define the fractal dimension of a HyperCosmos space of dimension r + 4 with

$$D(2^{r+4}) = \log(2^{r+4})/\log(2) = \log_2(2^{r+4}) = r + 4 \quad \text{for HyperCosmos Spaces} \qquad (2.1)$$

$$D(2^{r+3}) = \log(2^{r+3})/\log(2) = \log_2(2^{r+3}) = r + 3 \quad \text{for 2nd Kind HyperCosmos Spaces}$$

See Figs. 2.1, 2.2, and 2.3. In general we define

$$D(2^n) = \log(2^n)/\log(2) = \log_2(2^n) = n \qquad (2.2)$$

and an inverse

$$D^{-1}(n) = 2^n$$
$$D^{-1}(D(2^n)) = 2^n$$
$$D(D^{-1}(n)) = n$$

We identify the dimension array sizes as corresponding to fractal dimensions. (They are also the lengths of piecewise linear line segments in the construction of the Cosmos fractal curve.)[2]

If we "add" two spaces of dimensions r and r′, the combined space has the dimension array size:

HyperCosmos Space: $$d_{dr}\, d_{dr'} = 2^{r+4} * 2^{r'+4} = 2^{r+r'+8} = 16 d_{d(r+r')} \qquad (2.3)$$
2nd Kind HyperCosmos Space: $$d_{dr2}\, d_{dr'2} = 2^{r+3} * 2^{r'+3} = 2^{r+r'+6} = 8 d_{d(r+r')2}$$

The corresponding Hausdorff dimension "additions" give the combined dimensions as[3]

$$D(2^{r+4}) \text{ “+” } D(2^{r'+4}) = D(2^{r+r'+8}) \qquad (2.4)$$
$$D(2^{r+3}) \text{ “+” } D(2^{r'+3}) = D(2^{r+r'+6})$$

It is clear that dimension addition adds the powers of 2. We can interpret the "addition" in eq. 2.4 with blocks of dimension arrays as:

---

[22] These values change to 2^{r+7} and 2^{r+6} respectively when strata and fermion gambols are introduced. The results are similar with these choices.

[3] We have chosen to use base 2 in our study of dimensions because of its appearance in Cosmos spaces. Other choices of base are possible.

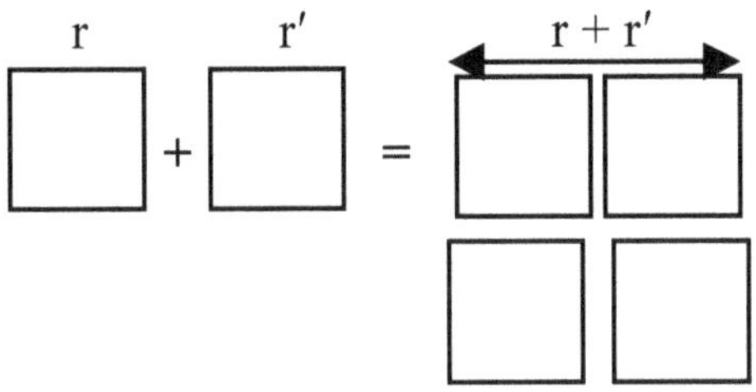

where the block sizes are multiplied by a factor of 4 or 3 as implied by eqs. 2.4.

2.1.2.2 Fractal Subtraction

Fractal subtraction causes removal of parts of a fractal. Again we view it in the context of the dimension arrays of Cosmos spaces.

$$D(2^{r+4}) \text{ “–” } D(2^{r'+4}) = D(2^{r-r'}) \qquad (2.5)$$
$$D(2^{r+3}) \text{ “–” } D(2^{r'+3}) = D(2^{r-r'})$$

based on

$$2^{r+4}2^{-(r'+4)} = 2^{r-r'} \qquad (2.6)$$

$$2^{r+3}2^{-(r'+3)} = 2^{r-r'}$$

It is clear that dimension subtraction subtracts the powers of 2. Subtraction may be viewed in terms of dimension arrays as

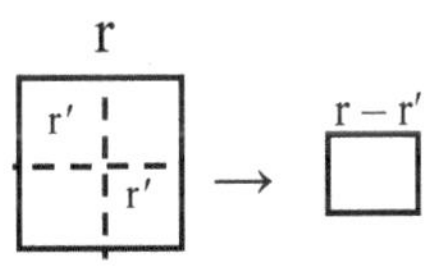

The interior r′ parts are excised by the subtraction. Fractal subtraction assumes $r > r'$.

2..1.2.3 Fractal Multiplication

Non-commutative Fractal multiplication can be implemented within the framework of Cosmos spaces:

$$D = D(2^{r+4}) \text{ “•” } D_C(2^{r'+4}) \qquad (2.7)$$

based on

$$2^{r+4} \wedge 2^{r'+4} = 2^{(r+4)k} \qquad (2.8)$$

where the ^ symbol indicates exponentiation. The constant k is

$$k = 2^{r'+4} \qquad (2.9)$$

Then

$$D = \log_2(2^{(r+4)k}) = k \log_2(2^{(r+4)}) = (r+4)2^{r'+4} \qquad (2.10)$$

For 2nd Kind HyperCosmos dimension arrays

$$D = \log_2(2^{(r+3)k'}) = k' \log_2(2^{(r+3)}) = (r+3)2^{r'+3} \tag{2.11}$$

wth $k' = 2^{r'+3}$.

Fractal multiplication for Cosmos spaces takes every element of a dimension array and expands it (through multiplication) by the dimension array of the other multiplicand.

2.1.2.4 Fractal Division

Fractal division can be implemented within the framework of Cosmos spaces:

$$D = D(2^{r+4}) \text{ “/” } D_C(2^{r'+4}) \tag{2.12}$$

based

$$2^{r+4} \wedge 2^{-(r'+4)} = 2^{(r+4)/k''} \tag{2.13}$$

where the ^ symbol indicates exponentiation. The constant k is

$$k'' = 2^{r'+4} \tag{2.14}$$

Then

$$D = \log_2(2^{(r+4)/k''}) = \log_2(2^{(r+4)})/k'' = (r+4)2^{-(r'+4)} \tag{2.15}$$

Similarly for 2nd Kind HyperCosmos dimension arrays:

$$D = (r+3)2^{-(r'+3)} \tag{2.16}$$

Fractal division of Cosmos spaces requires factoring within the r dimension array extracting r′ dimension array factors. Thus r > r′ is required or the result is a non-integer fractal dimension. Eqs. 2.15 and 2.16 explicitly show the factors.

2.1.2.5 Extension to Real Fractals

The addition of fractals in general combines the unit squares of two fractals in a manner similar to the illustration in section 2.1.2.1.

The subtraction of fractals in general separates the unit squares of the two fractals in a manner similar to the illustration in section 2.1.2.2. The terms in the fractal constructions must be separated to obtain the result. This presumes an extraction method (up to a multiple of the construction numbers) supporting their removal.

The multiplication of fractals in general combines the unit squares of two fractals in a manner similar to the description in section 2.1.2.3. The terms in one multiplicand of its fractal construction must be multiplied by the terms of the construction of the second multiplicand. The result is a combined fractal with terms being fractals of the second multiplicand.

The division of fractals in general decomposes the unit square of one fractal piecewise linear curve by piecewise linear curve extraction of the divisor fractal. The terms in the dividend in its fractal construction are divided by the terms of the construction of the divisor.

2.2 The Fractal Curve of the Cosmos – Double Threaded

The Cosmos Curve begins with a zero dimension point, and grows to a one dimension piecewise linear line segment by a simple aggregation of infinitesimal elements based on the identity

$$\sum_{k=\infty}^{1} 2^{-k} = 1 \tag{2.17}$$

It starts from a point where $k = \infty$, initially adding infinitesimal increments, which then sum to a unit line segment. The order of the summation begins with the infinitesimal and proceeds to the larger summands.

Then the unit piecewise linear line segment grows through a construction consisting of steps to form a fractal line. At the nth stage of the construction the growing line has the length 2^n. This line is compressed to fit within a unit square of sides 1 by 1. The line becomes fractal in the limit $n \to \infty$. The resulting fractal line is known as a Hilbert curve and, in the present context, the Cosmos Fractal Curve.

2.2.1 The Steps of the Construction

The steps of the construction process display a feature that relates to the distinction between HyperCosmos spaces and 2nd Kind HyperCosmos spaces.[4] The feature follows from our requirement that r be an even integer.

For *even* n we have the piecewise linear segment of length 2^n that corresponds to a HyperCosmos space of dimension r whose dimension array size is 2^{r+4}. We can map the linear segment length to the dimension array size by linearizing the dimension array square into a sequence of dimensions. The result is

$$2^n = 2^{r+4} \tag{2.18}$$

Then

$$n = r + 4 \tag{2.19}$$

In terms of dimensions we see

$$D(2^n) = D(2^{r+4}) \tag{2.19a}$$

for a HyperCosmos dimension array map.

For *odd* n we have a piecewise linear segment of length 2^n. In order to preserve the oddness of n we must relate it to a 2nd Kind HyperCosmos space of dimension r whose dimension array size is 2^{r+3}. Again we can map the linear segment length to the dimension array size by linearizing the dimension array into a sequence of dimensions. The result is

$$2^n = 2^{r+3} \tag{2.20}$$

---

[4] A HyperCosmos space has both a Normal and a Dark sector. A 2nd Kind HyperCosmos space only has a Normal sector. As a result their dimension array sizes differ by a factor of 2.

with

$$n = r + 3 \tag{2.21}$$

or

$$D(2^n) = D(2^{r+3}) \tag{2.21a}$$

for a 2nd Kind HyperCosmos space. The odd n is matched by the odd value of r+ 3 of 2nd Kind HyperCosmos dimension arrays since r is an even integer.

2.2.2 The Sequence of Construction Steps

Given the above comments, we can define the construction steps in terms of HyperCosmos spaces as:

$$D(2^n) = \delta_{n,even} D(2^{r+4}) + \delta_{n,odd} D(2^{r+3}) \tag{2.22}$$

where r +4 or r + 3 is the (topological or fractal) dimension of the corresponding Cosmos space

$$\begin{aligned} D(2^n) &= n \\ D(2^{r+4}) &= r + 4 \\ D(2^{r+3}) &= r + 3 \end{aligned} \tag{2.23}$$

The length n may also be viewed as a dimension of a space associated with the piecewise linear curve.

Eq. 2.22 shows that the construction steps of the fractal curve alternate between HyperCosmos space and 2nd Kind HyperCosmos space equivalents. We may view the construction process as alternating construction steps of HyperCosmos space, and 2nd Kind HyperCosmos space, piecewise linear segments.

Fig. 2.6 displays a part of the Cosmos Fractal Curve showing the "oscillation" between Cosmos and 2nd Kind Cosmos steps in n.

2.2.3 Extension of the Dimension Array Concept

By viewing a dimension array size as a fractal dimension, the possibility of extending our view of the Cosmos Curve appears. For each construction step n we can view the dimension array size[5] (2^{r+4} or 2^{r+3}) as the fractal (topological) dimension of its associated Cosmos space.

For each n a new second space also appears. This space exactly corresponds to the space-time dimension r. There is thus a coordinate space associated with each step n and its associated Cosmos space. One can view the Cosmos fractal construction process as also generating both a Cosmos space and a space-time at each step n.

The creation of a universe of the type of a Cosmos space r utilizes its associated space-time of dimension r. Cosmos spaces have both dimension arrays *and* space-times that are used in the formation of universes. A universe of space type r thus becomes a fractal as well and has the dimension r.

---

[5] At step n we have n = r + 4 for n even, and n = r + 3 for n odd.

2.3 Second Kind HyperCosmos Spaces

The appearance of 2nd Kind HyperCosmos space steps in the Cosmos fractal construction corresponds to their appearance in the unification of Cosmos spaces. Fig. 2.4 shows the set of unification spaces of Cosmos Theory. This figure and the following Fig. 2.5 were presented in previous books by the author.

Fig. 2.5 shows the dimension array of the 42 dimension Full HyperUnification space. It has three parts: an A part for HyperCosmos spaces unification, a B part for 2nd Part HyperCosmos spaces unification, and a C part for a remaining dimension array.

Dimension array A part is 5,592,400 ×5,592,400 dimensions
Dimension array B part is 2,796,200 ×2,796,200 dimensions
Dimension array C part is 8×8 dimensions

The 2nd Kind HyperCosmos spaces appear here (and in parts of previous books) that correspond directly with the structure of the Cosmos Fractal Curve presented above. Thus we have a consistency between the Cosmos Fractal Curve construction and the analysis of the unification structure of Cosmos Theory spaces.

2.4 Double Stranded Fractal Cosmos Curve and DNA

The sequence of construction of the Cosmos Fractal Curve alternates between HyperCosmos and 2nd Kind HyperCosmos space dimensions. The even and odd n construction steps support a separation of the Cosmos Curve into two strands that are interconnected step by step.

One can make an analogy of this construction with double helix DNA. This proposal strengthens the concept that Cosmos Theory is sufficiently complex as to justify being characterized as a type of organism – a Physical Organism – as opposed to an organic organism. See Fig. 2.6.

This chapter has shown the detailed relation of Cosmos Curve construction steps to HyperCosmos and 2nd Kind HyperCosmos spaces. It has also associated a space with (topological) dimension, and also a space-time for each construction step n.

THE HYPERCOSMOS SPACES SPECTRUM

Blaha Space Number $N = o_s$	Cayley-Dickson Number n	Cayley Number Cn d_c	Dimension Array column length d_{cd}	Dimension Array Size $d_{dN}=d_{dr}$	Space-time-Dimension r	CASe Group $su(2^{r/2},2^{r/2})$ CASe
0	10	1024	2048	2048^2	18	su(512,512)
1	9	512	1024	1024^2	16	su(256,256)
2	8	256	512	512^2	14	su(128,128)
3	7	128	256	256^2	12	su(64,64)
4	6	64	128	128^2	10	su(32,32)
5	5	32	64	64^2	8	su(16,16)
6	4	16	32	32^2	6	su(8,8)
7	**3**	**8**	**16**	$\mathbf{16^2}$	**4**	**su(4,4)**
8	2	4	8	8^2	2	su(2,2)
9	1	2	4	4^2	0	su(1,1)
10	0	1	2	2^2	-2	
11	-1	½	1	1	-4	

Figure 2.1. The HyperCosmos space spectrum augmented with N = 10 and N = 11 lines for use later. (Spaces with negative space-times may have universes. See Blaha (2022c).)

THE Limos SECTOR SPACES SPECTRUM

Blaha Space Number $N = O_s$	Cayley-Dickson Number n	Cayley Number d_c	Row Size d_{cr}	Dimension Array Size $d_{dN}=d_{dr}$	Space-time-Dimension r	CASe Group $su(2^{r/2},2^{r/2})$ CASe
10	0	1	2	2^2	-2	U(1)
11	-1	½	1	1^2	-4	U(½)
12	-2	¼	½	$½^2$	-6	U(¼)
13	-3	1/8	¼	$¼^2$	-8	U(1/8)
14	-4	1/16	1/8	$1/8^2$	-10	U(1/16)
				•		
				•		
				•		
∞	−∞	0	0	0	−∞	U(0)

Figure 2.2. The Cosmos Theory space spectrum Limos spaces for Gambol Theory. See Blaha (2023e).

HYPERCOSMOS OF THE SECOND KIND SPACES SPECTRUM

Blaha Space Number $N = O_s$	Cayley-Dickson Number n	Cayley Number d_c	Dimension Array size $d_{dN2} = d_{dr2}$	Space-time-Dimension r	CASe Group $su(2^{r/2}, 2^{r/2})$ CASe
0	10	1024	1024 × 2048	18	su(512,512)
1	9	512	512 × 1024	16	su(256,256)
2	8	256	256 × 512	14	su(128,128)
3	7	128	128 × 256	12	su(64,64)
4	6	64	64 × 128	10	su(32,32)
5	5	32	32 × 64	8	su(16,16)
6	4	16	16 × 32	6	su(8,8)
7	**3**	**8**	**8 × 16**	**4**	**su(4,4)**
8	**2**	4	4 × 8	2	su(2,2)
9	**1**	2	2 × 4	0	su(1,1)
10	0	1	1 × 2	-2	
11	-2	½	½	4	

Figure 2.3. Part of the HyperCosmos of the Second Kind space spectrum. (Spaces with negative space-times may have universes. Negative space-times indicate constraints.)

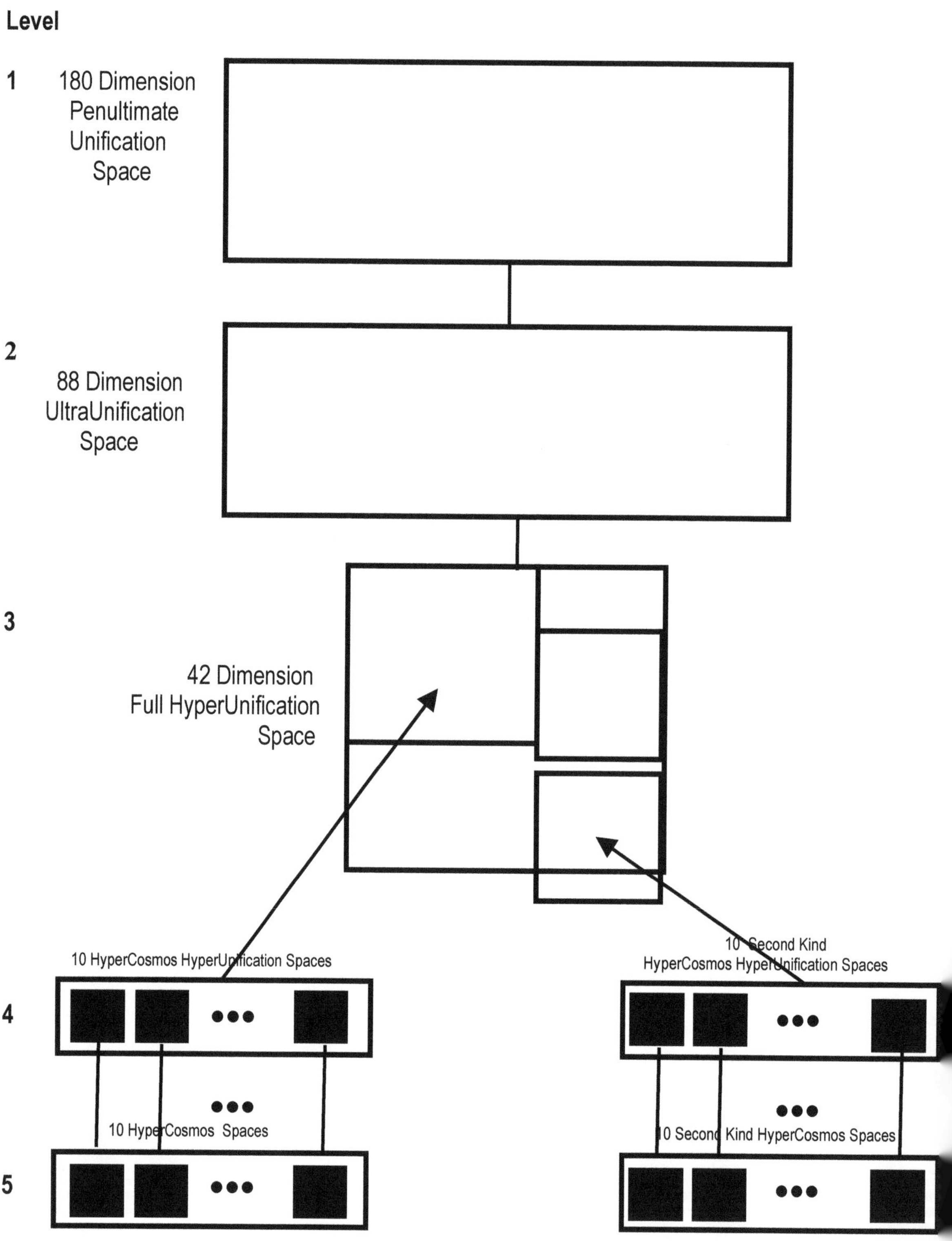

Figure 2.4. Diagram of the five levels of the Cosmos spaces unification. They contain 43 spaces.

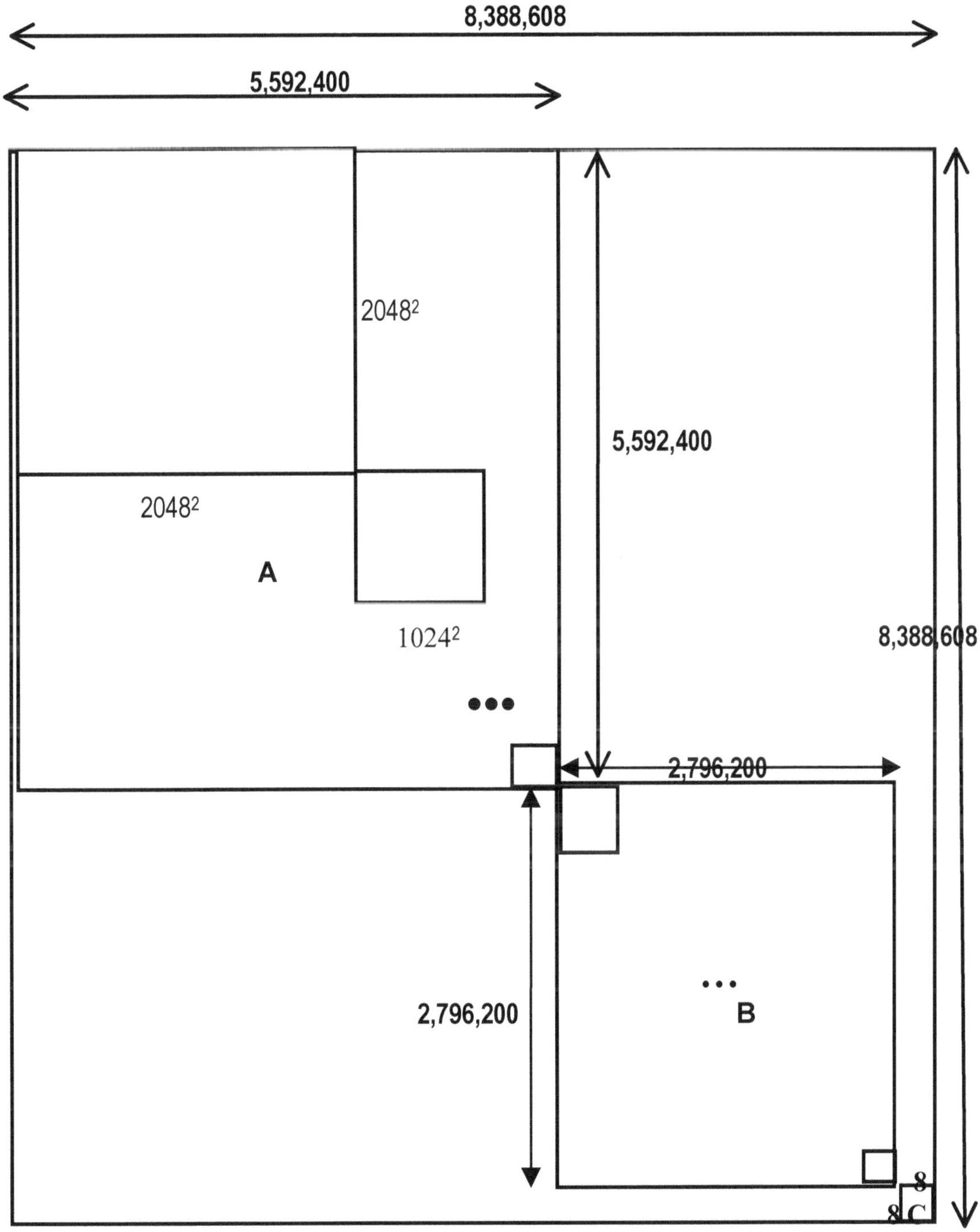

Figure 2.5. Form of the 42 space-time dimnesion of the Full HyperUnification space. All blocks are square. The figure is not drawn to scale.

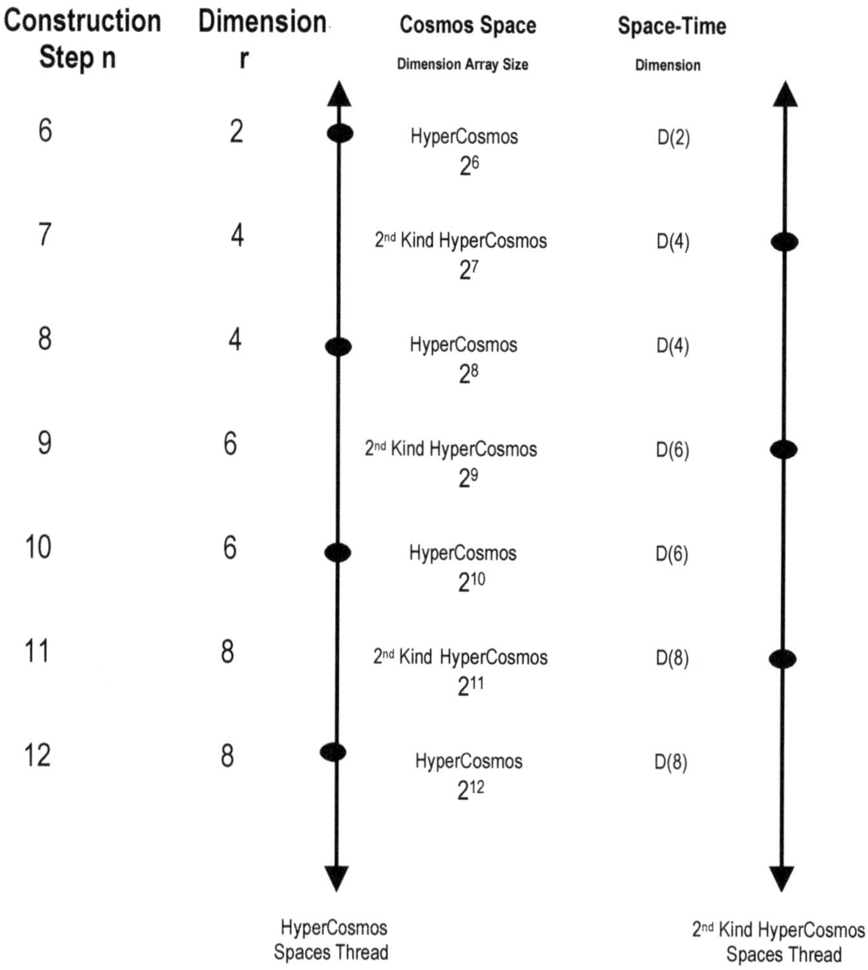

Figure 2.6. Section of the Cosmos Curve showing the two threads of Cosmos spaces and space-times. The space-times D(r) for each n is based on eq. 2.1. The piecewise linear line length is 2^n. At each construction step n there is the value of the linear line, its fractal equivalent, and the space-time equivalent. Note the repetitions of space-time dimensions is correct since it reflects the presence of both HyperCosmos and 2nd HyperCosmos spaces in the two threads. The "dots" indicate the type and presence of each Cosmos space on each thread for that value of n. Note their alternation between threads as n increases. The powers of 2 are the dimension array sizes.

3. Genesis of the Master Equation

The Cosmos Fractal curve has two significant features: a sequence of piecewise linear segments of length 2^n for construction step n, and the compression of these sequences to fit within a unit square. In the $n \to \infty$ limit the sequence becomes a space occupying fractal.

The sequence of piecewise linear segment lengths is 2^n. We can define a function that encapsulates this series of lengths in the form

$$\psi(u) = \cos(\pi \log_2(u)) \quad (3.1)$$

where u is a parameter. The zeros of $\psi(u)$ are at

$$u_n = 2^n \quad (3.2)$$

for integer n. The form of this solution and its zeros are scale independent – as it should be since no scale is present for the fractal construction.

We have seen in earlier books that eq. 3.1 is a special case of the type of the solutions of the Master Equation. The Master Equation is based on a scaling $1/r^2$ potential. The equation is a second order differential equation similar to the two dimension Schrödinger equation in a two dimension space.

The fractal Cosmos Curve has fractal dimension 2. It generates a unit square. This unit square may be scaled to an infinite two dimension space. The Master Equation is defined within the framework of this space.

It is a Schrödinger-like equation with a $1/r^2$ potential. The equation scales and allows us to avoid the introduction of a mass scale. The scaling $1/r^2$ potential has the advantages:

1. No mass scale dependence.
2. A spectrum of powers of a multiplier times an overall constant.
3. A constant that can be scaled out of the Master Equation by scaling the radial coordinate.
4. The capability to scale any energy eigenvalue to one.
5. The generality of the scaling $1/r^2$ Master Equation solution is supported by its solution also being equivalent to a harmonic oscillator-like solution and also a Free Field solution for small values of the radial parameter.

Perhaps the most significant feature of the 1/r2 potential is that it generates a set of eigenvalues in the form of a sequence of powers of a constant. This feature corresponds to the power series of the Cosmos spaces dimension arrays. See Fig. 2.6 where they are listed.

3.1 Master Equation Forms and Solutions[6]

The scaling potential provides the simplest guarantee of an energy spectrum that is a sequence of powers of a constant.[7] The Master Equation is:[8]

$$[(\partial^2/\partial x^2 + \partial^2/\partial y^2)/m + \lambda/m^2r^2 + C)\psi = 0 \tag{3.3}$$

where λ is a coupling constant and $r = (x^2 + y^2)^{½}$. In polar coordinates

$$x = u\cos(\varphi)$$
$$y = u\sin(\varphi)$$

we find

$$(\partial^2/\partial u^2 + 1/u\ \partial/\partial u + \partial^2/\partial\varphi^2 + \lambda/mu^2 + mC)\psi = 0 \tag{3.4}$$
$$(\partial^2/\partial\varphi^2)\psi = 0$$

and the solution

$$\psi = \psi(u)\Phi(\varphi) \tag{3.5}$$

where

$$\Phi(\varphi) = \exp(ik\varphi) \tag{3.6}$$

The quantum number k must be a whole integer to avoid irregularities in $\psi(r)$. We set $\lambda = k^2 + n_\varphi{}^2$. Substituting and separating factors in eq. 3.4 we find

$$(\partial^2/\partial u^2 + 1/u\ \partial/\partial u + (\lambda - k^2)/mu^2 + mC)\ \psi(u) = 0$$
$$(\partial^2/\partial u^2 + 1/u\ \partial/\partial u + \lambda_{n\varphi}u^2 + mC)\ \psi(u) = 0 \tag{3.7}$$

where[9]

$$\lambda_{n\varphi} = n_\varphi{}^2/m \tag{3.8}$$

$$C_{n\varphi n} = c_{n\varphi}\exp(n\pi/(\lambda_{n\varphi} - ¼)^{½}) \tag{3.9}$$

where n is the radial quantum number and n_φ is the angular momentum quantum number. The growth factor (multiplier) for each power series sequence is

$$g_{n\varphi} = \exp(\pi/(\lambda_{n\varphi} - ¼)^{½}) \tag{3.10}$$

---

[6] The discussions of these sections appear in more detail in Blaha (2025c).
[7] This equation may be viewed within the context of conformal symmetry due to its local scale invariant potential.
[8] This differential equation might be generalized to a fractal differential equation. However the coordinates are c-numbers. So a fractal extension seems unwarranted.
[9] The eigenvalue spectrum results from the imposition of orthogonality condition on the solution function near u = 0 according to von Neumann.

3.2 The Solutions of the Scaling Master Equation

The solution of eq. 3.7 of the Master Equation can be transformed to a function w(r) that makes the equation Schrödinger-like:

$$\psi(u) = u^{-1/2}\, w(u) \tag{3.11}$$

and obtain

$$(\partial^2/\partial u^2 + \lambda_{n\varphi}/mu^2 + mC)w(u) = 0 \tag{3.12}$$

where

$$\lambda_{n\varphi} = n_\varphi{}^2/m \tag{3.13}$$

and n_φ is a positive or negative integer. We now consider the solutions of eq. 3.12 for small r. The complete solution is a sum of Bessel functions.[10]

The solution for w(u) for small u is

$$w(u) \approx Br^{1/2} \cos((\lambda_{n\varphi} - 1/4)^{1/2}\ln u + b) \tag{3.14}$$

implying

$$\psi(u) \approx B\cos((\lambda_{n\varphi} - 1/4)^{1/2}\ln u + b) \tag{3.15}$$

where B and b are constants and ln() is the natural logarithm $\ln_2(u)$.Eq. 2.14 has a form similar to eq. 3.1 showing the close similarity of the zeros of the solutions of eq. 3.14 to the Cosmos spaces of the Cosmos Fractal Curve.

3.3 Zeros of the Master Equation Solution

The oscillating cosines above give us the set of zeros of the solution:

$$\begin{aligned}(\lambda_{n\varphi} - 1/4)^{1/2}\ln u + b &= \pm\pi/2,\ \pm 3\pi/2,\ \pm 5\pi\ 2,\ \pm 7\pi/2m\ ,,, \\ &= (n + 1/2)\,\pi\end{aligned} \tag{3.16}$$

where n is a positive or negative integer. Solving eq. 3.16 for r we find zeros at

$$u_n = \exp(((n + 1/2)\,\pi - b)/(\lambda_{n\varphi} - 1/4)^{1/2}) \tag{3.17}$$

If we set

$$b = \pi/2$$

and let n → –n, then

$$u_n = \exp(n\pi/(\lambda_{n\varphi} - 1/4)^{1/2}) = C_{n,n\varphi} \tag{3.18}$$

where $C_{n,n\varphi}$ is the eigenvalue of eq. 3.9 corresponding to the Cosmos dimension array sizes found in our analysis of the Cosmos Fractal curve.

---

[10] Since we can scale the r coordinate due to the scale covariant $1/r^2$ potential we can take the small r solution to be extended indefinitely.

3.4 Other Related Forms of the Master Equation

The Master Equation has other familiar forms when the parameter u is suitably transformed. It has a harmonic oscillator-like form and also a free wave function form.

If we define a variable s with

$$s = \ln u \tag{3.19}$$

then

$$(\partial^2/\partial s^2 - \partial/\partial s + \lambda_{n\varphi}/m - mCu^2)w(r) = 0 \tag{3.20}$$

Letting

$$w(u) = e^{s/2}\, v(s) = u^{1/2}\, v(s) \tag{3.21}$$

where, from eq. 2.10, we obtain

$$v(s) = u^{-1/2}\, w(u) = \psi(u) \tag{3.22}$$

Then we find

$$(\partial^2/\partial s^2 + (\lambda_{n\varphi}/m - 1/4) \quad - mCu^2)v(s) = 0 \tag{3.23}$$

or

$$(\partial^2/\partial s^2 + (\lambda_{n\varphi}/m - 1/4) \quad - mCe^{2s})v(s) = 0 \tag{3.24}$$

The mCe^{2s} term is the potential after the transformation to a harmonic oscillator-like form. We now connect the v(s) solution[11] for small u and thus large negative s to an approximate $\psi(u)$ solution for small u given in eq. 3.15. We approximate eq. 3.24 with

$$(\partial^2/\partial s^2 + \lambda_{n\varphi}/m - 1/4\,)v(s) = 0 \tag{3.25}$$

Eq. 3.25 is a free equation for v(s). The generality of the Master Equation solution is supported by its solution being equivalent to a Free Field solution (eq. 3.27 below) for small values of the radial parameter u.

A relevant form of the solution for comparison with eq. 3.15 is (using similar constants)

$$v(s) \approx B\cos((\lambda_{n\varphi} - 1/4)^{1/2}\, s + b) \tag{3.26}$$

Note eq. 3.15 can be expressed as

$$\psi(u) \approx B\cos((\lambda_{n\varphi} - 1/4)^{1/2} s + b) \tag{3.27}$$

Eq. 3.25 corresponds to a free quantum field. There is no direct quantization of the energy. The energy is a positive number. However we can impose a quantization based on a requirement of orthogonality of solutions (an observation due to von Neumann and

---

[11] We approximate the potential at its minimum relying on the scalability of u. Thus the small u (large negative s) neighborhood represents the potentials well.

Case) as we did in the scaling case. Then the energy is quantized due to the form of the solution in eqs. 3.26 and 3.27 for s. The quantized energy $E_{vn\varphi n}$ is

$$E_{vn\varphi n} = n\pi/(\lambda_{n\varphi} - ¼)^{½} \qquad (3.28)$$

It agrees with the scaling Master Equation quantization of eq. 3.9:

$$C_{n\varphi n} = c_{n\varphi} \exp(n\pi/(\lambda_{n\varphi} - ¼)^{½}) = c_{n\varphi} \exp(E_{vn\varphi n}) \qquad (3.29)$$

The constant n can be positive or negative. We choose n to be positive with n and n_φ being independent quantum numbers.

We thus find agreement between this approximate harmonic oscillator form of the Master Equation and the scaling Master Equation.

The transformation to an approximate harmonic oscillator form implies a form of stability for the Master Equation solutions. It also suggests a universality due to the generality of harmonic oscillator solutions under small perturbations for many Physical forces.

3.5 Equally Spaced Harmonic Oscillator-like Eigenvalues

The solution for ψ as a function of s displays equally spaced harmonic oscillator eigenvalues. From eq. 3.26 we find the zeros of ψ satisfy the analogue of eq. 3.16:

$$(\lambda_{n\varphi} - ¼)^{½} s_n + b = \pm\pi/2, \pm 3\pi/2, \pm 5\pi\ 2, \pm 7\pi/2m ,,, \qquad (3.30)$$
$$= (n + ½)\,\pi$$

or

$$s_n = ((n + ½)\,\pi - b)/(\lambda_{n\varphi} - ¼)^{½} \qquad (3.31)$$

Setting b = ½π results in an equally spaced Harmonic oscillator-like spectrum of zeros:

$$s_n = n\pi/(\lambda_{n\varphi} - ¼)^{½} \qquad (3.32)$$

with

$$r_n = \exp(s_n) \qquad (3.33)$$

which corresponds to eq. 3.18.

The approximate harmonic oscillator Master Equation implies the same spectrum and corresponding function zeros as the scaling Master Equation. The Master Equation has the desired scaling and fractal features. It is complemented by the harmonic oscillator-like equivalent.

We have found an approximate Harmonic Oscillator-like form and Free Field form of the Master Equation. It supports the generality of the Master Equation.

4. The Primeval Point - Master Equation Connection

The primeval Mathematical point leading to the Fractal unit square and to a two dimension space superficially might appear to be disjoint from the Master Equation. The Master Equation generates a progression of structure that leads fairly directly to coupling constant values and fundamental fermion masses.

At first glance the progression from Mathematical point to two dimension space does not appear to have experimental support. Thus there is a perceived conceptual "gap" between the progression and the Master Equation.

There are a number of reasons supporting the connection of the progression to the Master Equation:

1. The Master Equation solutions have zeros that can be put into a one to one correspondence with the piecewise linear line segments used in the construction of the fractal Cosmos curve that leads to the 2D space for the Master Equation. The values of $u_n = 2^n$ specify the length of the piecewise linear line segments that generate a Cosmos (and Hilbert) fractal curve. Other choices of the $g_{n\varphi}$ value generate other power sequences that can be scaled to the Hilbert sequence powers of 2^n. One can view the Master Equation as implying the fractal construction process. Thus the construction process from a point to the fractal-based unit square can be viewed as embodied in the Master Equation. The fractal construction process has an integral part in the generation of Cosmos structure.

2. The Master Equation solutions can be viewed as defining a generating function for the lengths of the piecewise linear line segments of the fractal.

3. The Master Equation has features that make it the simplest, and possibly the most general, choice of possible Cosmos generating equations:

    a. First it is a second order differential equation similar to the two dimension Schrödinger equation.
    b. It has a $1/u^2$ potential that enables the avoidance of the introduction of a mass scale.
    c. It can be transformed to a harmonic oscillator-like form with an equally spaced eigenvalue spectrum.
    d. The generality of the scaling Master Equation solution is supported by its solution also being equivalent to a Free Field solution for small values of the radial parameter.

4. The fractal curve construction supports a map between the 1D piecewise linear segments and the 2D unit square and space that preserves locality: two data points that are close to each other in a 1D line segment are close to each other after the line is folded in the fractal construction process. Locality is a very desirable feature in Physics.

5. The Master Equation, which embodies scaling, can be transformed to a harmonic oscillator-like form with a corresponding eigenvalue spectrum. This form, which characterizes small perturbative forces, provides a type of stability against disturbances from small forces.

    a. The generality of the scaling Master Equation solution is supported by its solution also being equivalent to a Free Field solution for small values of the radial parameter u.

6. We have shown from the Master Equation solution that the evolutionary sequence of the Cosmos follows. We see it encompasses the initial mathematical point, the generation of a two dimension piecewise linear unit length, the generation of the fractal Cosmos Curve, the generation of the Cosmos spaces and their dimension arrays, the unification spaces sequence, and the set of sequences of Fig. 5.2.

The Master Equation is the basis of the evolution of the Cosmos conceptually. As we have often pointed out, Cosmos evolution is instantaneous in the sense that there is no time coordinate. Therefore it is timeless! The Master Equation embodies the Cosmos and sets the stage for its universes. The unity of the progression from point to Master Equation is established.

5. Fractal Representation of Cosmos Space Sequences

In chapter 2 we developed a fractal view of the construction steps in the definition of the Cosmos Fractal Curve. At each step n we associated a fractal with a dimension array of $2^n = 2^{r+4}$ or 2^{r+3} of a HyperCosmos space or a 2nd Kind HyperCosmos space respectively. As one increases the value of r by 2 (since r values are even integers) the dimension array size increases by a factor of 4. We can represent this change with the equation:

$$2^{r+4}2^2 = 2^{r+6} = 2^{(r+2)+4}$$

or

$$D(2^{r+4}) + 4 = D(2^{r+2+4}) \quad \text{HyperCosmos space} \tag{5.1}$$

$$D(2^{r+3}) + 4 = D(2^{r+2+3}) \quad 2^{nd} \text{ Kind HyperCosmos space} \tag{5.2}$$

We now assume the quantity 4 that appears in these equations is a fractal dimension:

$$D(2^4) = 4 \tag{5.3}$$

by eq 2.2. We represent eqs. 5.1 and 5.2 as fractal additions:

$$\begin{aligned} D(2^{r+4}) \text{ “+” } D(2^4) &= D(2^{r+2+4}) \\ D(2^{r+3}) \text{ “+” } D(2^4) &= D(2^{r+2+3}) \end{aligned} \tag{5.4}$$

as described in chapter 2 in eqs. 2.1 and 5.4. Diagramatically we represent eq. 5.4 as

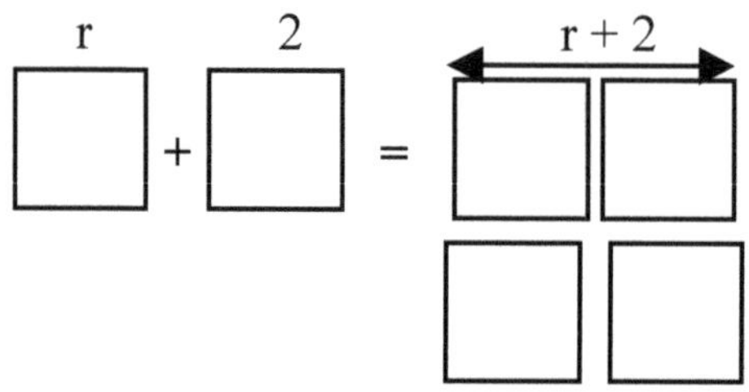

Figure 5.1. Addition of D(4) to a Cosmos Space Fractal. Note the D(4) signifies four items in a square array as shown with sides of 2. The D(4) indicates a dimension-type array of 4 entries and a row length of 2.

The r + 2 dimension array is generated. $D(2^4)$ represents 2^4 as a fractal using eq. 2.2:

$$D(2^4) = \log(2^4)/\log(2) = \log_2(16) = 4 \tag{5.5}$$

Eq. 5.4 shows the fractal $D(2^4)$ represents a multiplier of 4 for the dimension array for r extended to r + 2.

5.1 Dimension Array Column Length

If one extends the dimension array row (and column) length by 2 dimensions then one has the fractal addition:

$$D(r) \text{ “+” } D(2^2) = D(r + 2) \qquad (5.6)$$

which corresponds to the multiplier “2” in the A1 row of Fig. 5.2 for column lengths:

A1 Cosmos spaces dimension array column lengths $2^{r/2+2}$	4	$\mathbf{2} = D(4) = \mathbf{ln_2(4)}$	20.79

5.2 The Multiplier is a Fractal

The appearance of fractal addition in eqs. 5.4 suggests that the addend $D(2^4)$ is also a fractal dimension. When we consider the construction steps for the Cosmos Fractal Curve we find the fractal Cosmos space for dimension r increases to the fractal Cosmos space of dimension r + 2 for each construction step.

In particular, Fig. 2.6 shows that the steps for even values of n correspond to Cosmos spaces increase by 2, and also the steps for odd values of n correspond to 2nd Kind Cosmos spaces increase by 2.

The increase for Cosmos space steps (even n) may be viewed as

$$D(2^n) \text{ “+” } D(2^2) = D(2^{r+4}) \text{ “+” } D(2^2) = D(2^{r+6}) = D(2^{n+2}) \qquad (5.7)$$

based on n = r + 4 and

$$2^n 2^2 = 2^{r+4} 2^2 = 2^{(r+2)+4} = 2^{r+6} = 2^{n+2}$$

For odd n

$$D(2^n) \text{ “+” } D(2^2) = D(2^{r+3}) \text{ “+” } D(2^2) = D(2^{r+5}) = D(2^{n+2}) \qquad (5.8)$$

based on n = r + 3 and

$$2^n 2^2 = 2^{r+3} 2^2 = 2^{r+5} = 2^{(r+2)+3} = 2^{n+2}$$

Thus we can see that the increments $D(2^2)$, which correspond to multipliers of 4 for Cosmos dimensions r, have a fractal formulation. Thus we find multipliers for the Cosmos Fractal Curve are fractal.

Descending the construction steps is possible through the use of fractal subtraction:

$$D(2^n) \text{ “–” } D(2^2) = D(2^{r+4}) \text{ “–” } D(2^2) = D(2^{r+2}) = D(2^{n-2}) \qquad (5.9)$$

$$D(2^n) \text{ “–” } D(2^2) = D(2^{r+3}) \text{ “–” } D(2^2) = D(2^{r+1}) = D(2^{n-2}) \qquad (5.10)$$

for the HyperCosmos (even n) and 2nd Kind HyperCosmos (odd n) cases respectively.

5.3 All Multipliers are Fractal

The fractal nature of the Cosmos Fractal Curve multipliers seen above leads us to state all multipliers in Fig. 5.2 are fractal. We indicate this fact in Fig. 5.2 by showing the fractal logarithm value for each multiplier.

5.4 Extension to Coupling Constants and Fermions

In later chapters we will show that coupling constant and fermion mass sequences, which use Fig. 5.2 multipliers are also necessarily fractal. Their fractal nature raises profound questions of the origin of their fractality.

Sequence Fractal	n_φ	g_g Fractal Dimension multiplier factor	λ_{n_φ}
Fractal	12	$1{,}26 = \ln_3(4)$	187.1
Fractal	10	$1.317 = \ln_{2.7}(3.7)$ $= \ln_5(8)$	129.9
Fractal	8	$1.41 = \ln_{2.36}(3.36)$ $= \ln_4(7)$	83.16
Sierpinski Triangles Fractal; 1.58-bit Large Language Model	6	$1.585 = \ln_2(3)$	46.778
A1 Cosmos spaces dimension array column lengths $2^{r/2+2}$	4	$2 = \ln_2(4)$	20.79
A2 Cosmos unification space dimension r	4	$2 = \ln_2(4)$	20.79
A3 The sequence of Unification Spaces	4	$2 = \ln_2(4)$	20.79
B Coupling Constants α Multiplier	2	$4 = \ln_2(16)$	5.39
C Fermion Sequence 2 Masses	1	$2/3 \times 2^5 = \ln_2((32^{32})^{0.133})$	1.3
D Fermion Sequence 1 Masses	$½^{½}$	$3/2 \times \pi 2^5 = \ln_2((32^{32})^{0.3\pi})$	0.65
E n_φ Values Sequence; A Sequence of Sequences	½	$95939 = 5.96 \ln_2(32^{32\cdot 32\pi})$	0.325
F New n_φ Values Sequence based on α_{QED}	$0.48 \approx 2^{½}/3$	$844242 = 5.15\ln_2(32^{32\cdot 32\cdot 32})$	0.303

Figure 5.2. Extended Cosmos Theory Sequences with a sequence determined by setting λ = 0.303, which is the value of the QED couping constant e which determines the Fine Structure Constant α. The first row is for the case of a multiplier √2. Its n_φ value is an extension of the series of n_φ values 2, 4, 8 of rows A1 through B. The first and second rows are for the cases $n_\varphi = 6$ and $n_\varphi = 8$ which also may be of interest. The second row may correspond to 1.58-bit Large Language Models. Note: $32\cdot 32 = 32^2$ and $32\cdot 32\cdot 32 = 32^3$ above.

6. Fractal Representation of Coupling Constants

The sequence of coupling constants appears in Fig. 6.1. The coupling constants α_i have a clear multiplier – a factor of 2 – corresponding to row B of Fig. 5.2: Their multiplier is a fractal number – a fractal (topological in this case) dimension.

B Coupling Constants α Multiplier	2	$4 = \ln_2(16)$	5.39

Consistency requires the values of coupling constants α_i are also fractal numbers. The multiplicative[12] relation between the members of the sequence is:

$$D(2^{\alpha_i}) \text{ “•” } D(4) = D(2^{\alpha_{i+1}}) \tag{6.1}$$

based on

$$2^{\alpha_i} \wedge 4 = 2^{\alpha_i k} \tag{6.2}$$

where ^ indicates exponentiation and

$$k = 4 = 2^2 \tag{6.3}$$

Then

$$\log_2(2^{\alpha_i k}) = 2^{\alpha_{i+1}} \tag{6.4}$$

yields eq. 6.1.

The coupling constants also satisfy fractal division:

$$D(2^{\alpha_i}) \text{ “/” } D(4) = D(2^{\alpha_{i-1}}) \tag{6.1'}$$

based on

$$2^{\alpha_i} \text{ “/” } 4 = 2^{\alpha_i/k} \tag{6.2'}$$

We view the multipliers between α_i values as D(4) fractal dimensions in themselves. Then the actual α_i values are also fractal dimensions. *Thus there are six coupling constant fractals at present.*

[1212] See eqs. 2.9 and 2.10 for the general case of multiplication.

ESTIMATE				EXPERIMENT		
Interaction		Expression	$\alpha_i = g^2/4\pi$ Value	g	Known[13] Value $g^2/4\pi$	Deviation
U(0)	α_0	$e^2/4096$	0.0018036	-	-	-
U(1)	$\alpha_1 = \alpha$	$e^2/1024$	0.00721438	0.303	0.0072973525643	1.15%
SU(2)	$\alpha_2 = g^2/4\pi$	$e^2/256$	0.0289	0.63	0.0316	9.3%
SU(3)	$\alpha_3 = \alpha_S$	$e^2/64$	0.115	1.21	0.117	1.7%
SU(4)[14]	α_4	$e^2/16$	0.462	2.4?	0.458	0.087%

Figure 6.1. Coupling constants for the UST (and the Standard Model). Note factor of four relation between adjoining α values.[15]

6.1 The Coupling Constant Algorithm

The coupling constant counting algorithm is described in Blaha (2018e) and (2024i). Its form is consistent with the fractal representation with the multiplier 4 described above. The algorithm is:

> We have shown coupling constants appear to form a powers sequence. We now consider the expression of a coupling constant as the product of a number of factors that are based on geometry.
>
> First we note that n is the number of fermions in a fundamental representation of SU(n). The coupling constant $g(n) = g_n$ has the form:
>
> $$\begin{aligned} g(n) &= e\,(4\pi)^{1/2}\, 2^{n-6} = e\,(2^{-10}\pi)^{1/2}\, 2^n = 0.1505\ 2^n \\ &= g_g\,(\text{number of spin states per fermion})^{\text{number of fermions}} \\ &= g_g\,(\text{number of spin states per fermion})^n \end{aligned} \tag{1.18}$$
>
> n is the number of fermions in irreducible representation
>
> $$= g_g\, 2^n$$
>
> The coupling constant is a fundamental value g_g times the number of spin states (a geometric property of fermions equal to half the column length of a spinor) raised to the number of fermions. The number of spin states per fermion is 2 in r = 4 dimensions.
>
> We see that g(n) has n factors – one factor for 2 spin states for each fermion in the SU(n) fundamental representation. The origin of g_g appears to be
>
> $$g_g = e\,(2^{-10}\pi)^{1/2} \tag{1.19}$$
>
> Note
>
> $$\begin{aligned} \alpha_n &= g_n^{\,2}/4\pi = e^2\, 2^{2n-12} \\ &= \alpha_0\, 2^{2n} \end{aligned} \tag{1.20}$$

[13] All coupling constant values are based on data from Particle Data Group Tables of 2024.
[14] This value is based on the "doubling trend" seen in the three known coupling constants g_i above in Fig. 3.2.
[15] From Blaha (2018e).

Eq. 1.18 is *initially* not decisively understood due to the appearance of the natural logarithm base e. However e and π appear naturally later in our studies of higher dimension volumes and surface areas. We therefore view g_g as having a *geometric origin.*

Eq. 1.18 implies that the coupling g(n) for SU(n) is g_g times n factors of 2 which represents the number of spin states per fermion in the SU(n) fundamental representation in four dimensions. Thus it is a product of degrees of freedom.

An important point is the number of factors is equal to the number of fermions in the SU(n) irreducible representation. *The coupling constant "felt" by one fermion depends on the total number of fermions in the group's fundamental representation.*

6.2 Fractal Representation of the Coupling Constant Algorithm

The above algorithm may be simply expressed in terms of fractals. We begin by defining a fractal for SU(n) equal to 2 raised to the number of real-valued dimensions of the irreducible representation of the group:

$$D(SU(n)) = 2n \tag{6.5}$$

Then eq. 1.20 above has the fractal representation

$$D(2^{\alpha_i}) \text{ "•" } D(SU(n)) = D(2^{\alpha_{i+1}}) \tag{6.6}$$

giving

$$\alpha_n = \alpha_0 2^{2n}$$

7. Fractal Representation of the Coupling Constants Related to Fundamental Fermion Masses

This chapter relates fractal representation Coupling Constants to Fundamental Representation Masses. In doing this, we will see that the fractal nature of Coupling Constants carries over to the multipliers and λ values for fermion mass sequences. In particular the λ values and the multipliers for the mass sequences (rows C and D of Fig. 5.2) are directly related to Coupling Constants.[16]

We conclude the fractal nature of Coupling Constants and these relations to mass sequences implies that the multipliers and λ values for fermion mass sequences are also *fractal*. Therefore fermion masses are fractal.

7.1 Cosmos Coupling Constant Values Related to Fundamental Fermion Masses

This section relates the coupling constant sequence to the fermion mass sequences.

7.1.1 The Fermion Masses Sequences

In Blaha (2024g), we found that the eight known fundamental fermions of the first layer of the four dimension Cosmos Theory space and UST could be separated into two sequences of four fermions. Sequence 2 which contained the down-type fermions was approximately estimated to have a series of ratios of $2^5 = 32$ factors between the masses. Sequence 1 fermions (up-type) had less consistent factors between masses. We tentatively estimated the ratio $2^5\pi = 32\pi$ based on an approximation to the t/c fermion mass ratio and to achieve a rough parallel with the down-type fermion factor.

The actual mass values of sequence 1 deviated significantly from the $2^5\pi$ multiplier. We attribute the deviations to additional perturbative interaction effects on masses that are beyond this model.

7.1.2 The Fundamental Fermion Masses Model

The mass sequences, as pictured below in Fig. 8.1, have a geometric character similar to the energy levels and the Cosmos spaces spectrum.

We now define a 2×2 Master Equation scaling equation for *both* sequences:

$$\begin{bmatrix} \partial^2/\partial u^2 + \lambda_1/u^2 - M_1 & 0 \\ 0 & \partial^2/\partial u^2 + \lambda_2/u^2 - M_2 \end{bmatrix} \begin{bmatrix} \varphi_1 \\ \varphi_2 \end{bmatrix} = 0 \tag{7.1}$$

which leads to

---

[16] See Blaha (2025d) for more detail.

$$2^5\pi = \exp(\pi/(\lambda_1 - ¼)^{½}) \quad (7.2)$$
$$2^5 = \exp(\pi/(\lambda_2 - ¼)^{½}) \quad (7.3)$$

multipliers for the up-type and down-type sequences respectively seen in Fig. 5.2. Then

$$[(\ln(32\pi)/\pi]^{-2} + ¼ = \ = \lambda_1 = 0.714 \qquad \approx g_2 = 0.63 \quad (7.4)$$

$$[(\ln(32)/\pi]^{-2} + ¼ = \ = \lambda_2 = 1.07 \qquad \approx g_3 = 1.21 \quad (7.5)$$

where we adjoin Coupling Constant g values of comparable value from the coupling constant table (Fig. 6.1).

Note the values of λ_1 and λ_2 are related by:

$$1.07/.714 = 1.5 \quad (7.6)$$

Note the sum of the values $\lambda_1 + \lambda_2 = 1.784$ and also $g_2 + g_3 = 1.84$. These values differ by 3.4% suggesting that the ElectroWeak SU(2) and Strong SU(3) interactions combine to generate each mass sequence through an interplay of interactions.

7.1.3 Separation of SU(2) and SU(3) Contributions to the Sequences' Powers

We view the values in the sequences as sums of SU(2) and SU(3) terms. We calculate the discrepancy x between the λ_i from

$$\lambda_1 + x = 1.21 \quad (7.7)$$
$$\lambda_2 - x = 0.63$$

using the approximate equality: $\lambda_1 + \lambda_2 = 1.784$ while $g_2 + g_3 = 1.84$. We find

$$x = 0.468 \quad (7.8)$$

.

After noting that $g_4 = 2.4$ is itself an extrapolation we will use the approximate values from Fig. 5.2 to obtain

$$x = ½(g_1 + g_2) = 0.466 \quad (7.9)$$

Then we find an approximate relation between the SU(1), SU(2) and SU(3) coupling constants, and the λ_i, factors in in the fundamental fermion mass sequences.

$$\lambda_1 = 0.714 = 1.21 - 0.468 = g_3 - ½(g_1 + g_2) \quad (7.10)$$

for sequence 1, and

$$\lambda_2 = 1.07 = 0.63 + 0.468 = 3g_2/2 + ½ g_1 \quad (7.11)$$

for sequence 2.

Sequence 2 is governed by the ElectroWeak interactions. Sequence 1 is governed by a combination of ElectroWeak and Strong SU(3) interactions. This chapter directly relates fermion mass sequences to the coupling constant sequence.

As a result we find the fermion masses and their sequence λ values are also fractal due to their relation to Coupling Constant fractals.

7.2 Strata Fermion Masses are Fractal Coupling Constant Dependent

Each fundamental fermion is one of four fermion analogues that differ only in mass. The analogues have identical features except for their masses and their stratum number. The masses differ by a factor of 95939 according to the entry in Fig. 5.2. The masses of each set of four corresponding fermions must be fractal based on the above arguments for fundamental fermion mass fractals due to their relation to Coupling Constant fractals and the 95939 multiplier fractal in .Fig. 5.2.

7.3 Gambol Fermion Masses are Fractal

Each fundamental fermion has a corresponding gambol fermion with the same quantum numbers except for their stratum number and gambol number. Their masses are a factor of 1/844242 of their corresponding fermion. This mass relation implies that they also are the result of the same process that generates the fundamental fermions' mass seen above. Therefore we apply the same logic as in section 7.1 to relate the gambol masses to the SU(3), SU(2) and U(1) coupling constants – thus making them fractal as well.

We conclude all fermion masses are fractal their λ values are fractal, and 95939 and 844242 are fractal.

8. Fractal Representation of Fermion Masses

We have seen that fermion masses scaled by some mass scale have fractal dimensions. We now consider the approximate mass sequences of Fig. 8.1 treated as fractal and derive their fractal interrelations.

The approximate masses for the two fundamental fermion sequences appear in Fig. 8.1. The four strata masses of each fundamental fermion type differ by a factor of 95939. The gambol fermion masses are lower by a factor of 844242 than their corresponding fermion's mass.

	SEQUENCE 1				SEQUENCE 2			
	e	u	c	t	ν'	d	s	b
Mass:	0.511×10^{-3}	1.80×10^{-3}	1.28	171	8.4×10^{-5}	4.24×10^{-3}	102×10^{-3}	4.34
Multiplier:	$2^5\pi$	$2^5\pi$	$2^5\pi$		32	32	32	

Figure 8.1 The fundamental fermion mass sequences with approximate multiplier factors between the masses of each sequence. The sequence 1 masses show a significant variation in their $2^5\pi$ factor but their overall trend is approximately satisfactory. Sequence 2 masses better approximate the factor of 32 progression.

The relations between the masses in Sequence 1 is multiplicative. When we consider their fractal dimensions, we see the logarithmic nature of dimensions causes the dimensions to undergo multiplication in analogy to the Cosmos Fractal Curve construction steps in chapter 5. The fractal multiplier of $2^5\pi$ for sequence 1 takes a fermion mass to the next mass. We introduce a mass scale m_0 since fractal dimensions are dimensionless. We define the sequence 1 fractal mass ratios m_i

$$m_e/m_0,\ m_u/m_0,\ m_c/m_0,\ m_t/m_0$$

denoting them with an index i. Then

$$D(2^{m_i})\ \text{“}\bullet\text{”}\ D(2^{32\pi}) = D(2^{m_{i+1}}) \tag{8.1}$$

based on

$$2^{m_i} \wedge 2^k = 2^{m_i k} = 2^{m_{i+1}}$$

where

$$k = 2^5\pi = 32\,\pi$$

Similarly we find sequence 2 fractal mass ratios m_i

$$m_{\nu'}/m_0,\ m_d/m_0,\ m_s/m_0,\ m_b/m_0$$

also denoted with an index i gives

$$D(2^{m_i}) \text{ “•” } D(2^{32}) = D(2^{m_{i+1}}) \qquad (8.2)$$

based on

$$2^{m_i} \wedge 2^{k'} = 2^{m_i k'} = 2^{m_{i+1}}$$

where

$$k' = 2^5$$

The sequences are now fractal in nature. Each mass is fractal. The multipliers are fractal.

8.1 Stratum Fractals

The strata fermion masses are separated by a factor of 95939 as shown in Fig. 8.2. Their fractal dimensions satisfy the fractal addition as one ascends the strata

$$D(2^{m_i}) \text{ “•” } D(2^{95939}) = D(m_{i+1}) \qquad (8.2)$$

where the index i labels the strata in the order of increasing masses.

8.2 Gambol Fractals

The gambol fermion masses are separated from their corresponding fermion by a factor of 844242 as shown in Fig. 8.2. Their fractal dimensions satisfy the fractal division:

$$D(2^{m_{ifermion}} \text{ “/” } D(2^{844242}) = D(2^{m_{igambol}}) \qquad (8.3)$$

The preceding discussions lead us to propose that, not just masses, but the entire fermion particles are fractal and have a fractal substructure reflecting a deep inner dynamics.

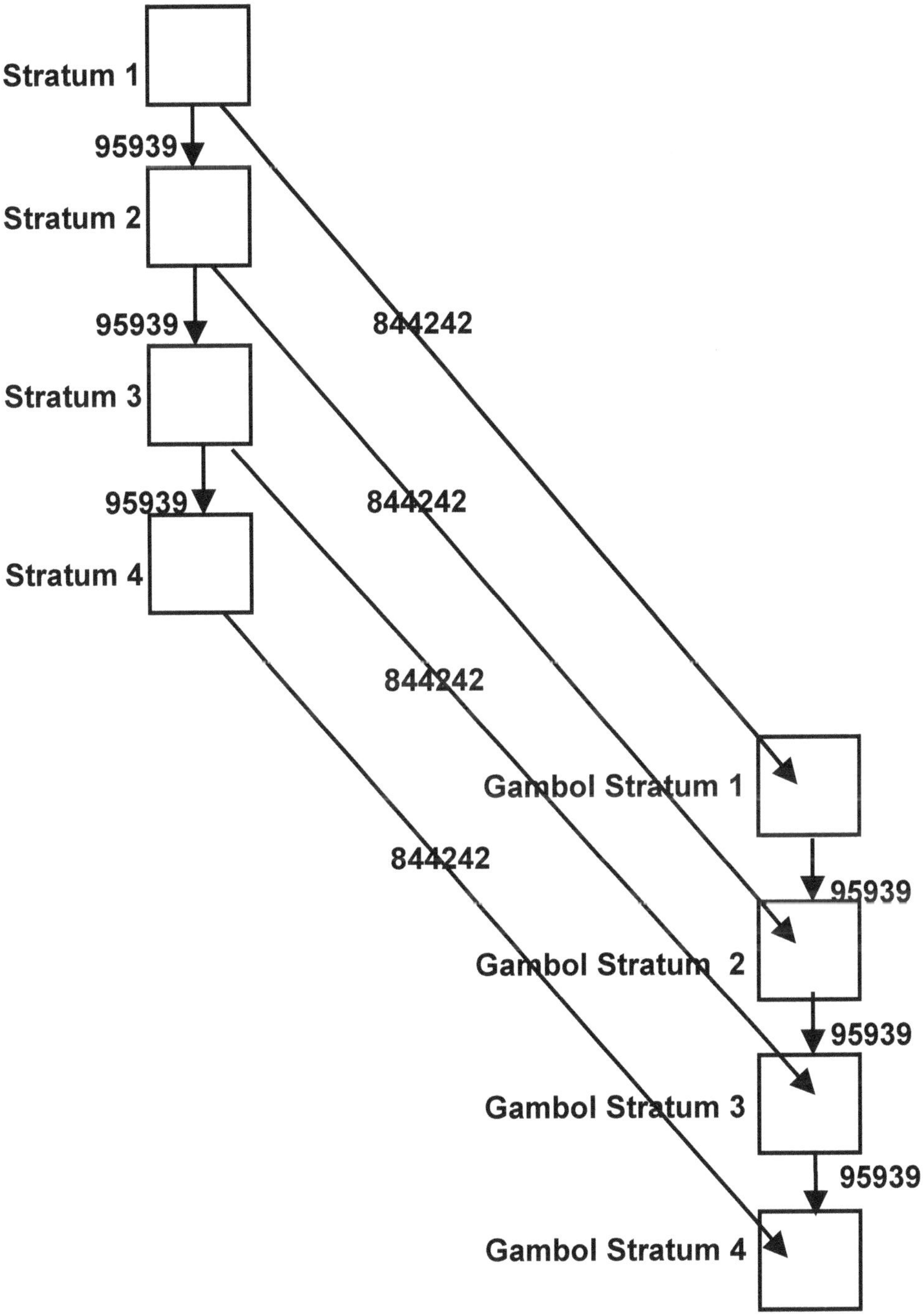

Figure 8.2. Each left block represents a 256 fermion set. Each right block represents a corresponding 224 gambol fermion set since neutrinos do not have gambols. Coresponding fermions in adjacent strata differ in mass by a factor of 95939 in increasing order from stratum 1. Each gambol fermion is a factor of 1/844224 of the corresponding fermion mass. Each gambol mass is 1/95939 of the mass of the gambol immediately above it. Not drawn to scale. All fundamental fermion particles pictured above are fractal.

9. Fractal Representations of Spaces, Particles and Interactions

We have developed the fractal representation of spaces, fermions and interactions. These components of Physical Reality are now revealed to have a deeper basis. In a sense the fractal representation – based as it is on scaling – raises the possibility of a direct relation to UST Quantum Field Theory and possibly the Renormalization Group.

The Cosmos Theory that we developed previously is a *simulation* of Reality. This new formulation of Cosmos Theory, based on a fractal formulation, is a much deeper *simulation* of Reality. We will return to the analysis of this new simulation in a future work.

We have come to an understanding of key points in the fractal construction of a theory of the Cosmos and its description as a Physical Reality.

REFERENCES

Akhiezer, N. I., Frink, A. H. (tr), 1962, *The Calculus of Variations* (Blaisdell Publishing, New York, 1962).
Bjorken, J. D., Drell, S. D., 1964, *Relativistic Quantum Mechanics* (McGraw-Hill, New York, 1965).
Bjorken, J. D., Drell, S. D., 1965, *Relativistic Quantum Fields* (McGraw-Hill, New York, 1965).
Blaha, S., 1995, *C++ for Professional Programming* (International Thomson Publishing, Boston, 1995).
_______, 1998, *Cosmos and Consciousness* (Pingree-Hill Publishing, Auburn, NH, 1998 and 2002).
_______, 2002, *A Finite Unified Quantum Field Theory of the Elementary Particle Standard Model and Quantum Gravity Based on New Quantum Dimensions™ & a New Paradigm in the Calculus of Variations* (Pingree-Hill Publishing, Auburn, NH, 2002).
_______, 2004, *Quantum Big Bang Cosmology: Complex Space-time General Relativity, Quantum Coordinates™ Dodecahedral Universe, Inflation, and New Spin 0, ½, 1 & 2 Tachyons & Imagyons* (Pingree-Hill Publishing, Auburn, NH, 2004).
______, *2005a, Quantum Theory of the Third Kind: A New Type of Divergence-free Quantum Field Theory Supporting a Unified Standard Model of Elementary Particles and Quantum Gravity based on a New Method in the Calculus of Variations* (Pingree-Hill Publishing, Auburn, NH, 2005).
______, 2005b, *The Metatheory of Physics Theories, and the Theory of Everything as a Quantum Computer Language* (Pingree-Hill Publishing, Auburn, NH, 2005).
______, 2005c, *The Equivalence of Elementary Particle Theories and Computer Languages: Quantum Computers, Turing Machines, Standard Model, Superstring Theory, and a Proof that Gödel's Theorem Implies Nature Must Be Quantum* (Pingree-Hill Publishing, Auburn, NH, 2005).
______, 2006a, *The Foundation of the Forces of Nature* (Pingree-Hill Publishing, Auburn, NH, 2006).
______, 2006b, *A Derivation of ElectroWeak Theory based on an Extension of Special Relativity; Black Hole Tachyons; & Tachyons of Any Spin.* (Pingree-Hill Publishing, Auburn, NH, 2006).
_______, 2007a, *Physics Beyond the Light Barrier: The Source of Parity Violation, Tachyons, and A Derivation of Standard Model Features* (Pingree-Hill Publishing, Auburn, NH, 2007).
______, 2007b, *The Origin of the Standard Model: The Genesis of Four Quark and Lepton Species, Parity Violation, the ElectroWeak Sector, Color SU(3), Three Visible Generations of Fermions, and One Generation of Dark Matter with Dark Energy* (Pingree-Hill Publishing, Auburn, NH, 2007).
______, *2008a, A Direct Derivation of the Form of the Standard Model From GL(16) (Pingree-Hill Publishing, Auburn, NH, 2008).*
______, 2008b, *A Complete Derivation of the Form of the Standard Model With a New Method to Generate Particle Masses Second Edition* (Pingree-Hill Publishing, Auburn, NH, 2008)
______, 2009, *The Algebra of Thought & Reality: The Mathematical Basis for Plato's Theory of Ideas, and Reality Extended to Include A Priori Observers and Space-Time Second Edition* (Pingree-Hill Publishing, Auburn, NH, 2009).
______, 2009a, *Bright Stars, Bright Universe* (Pingree-Hill Publishing, Auburn, NH, 2008)
______, 2010a, *Operator Metaphysics: A New Metaphysics Based on a New Operator Logic and a New Quantum Operator Logic that Lead to a Mathematical Basis for Plato's Theory of Ideas and Reality* (Pingree-Hill Publishing, Auburn, NH, 2010).
______, 2010b, *The Standard Model's Form Derived from Operator Logic, Superluminal Transformations and GL(16)* (Pingree-Hill Publishing, Auburn, NH, 2010).
______, 2010c, *SuperCivilizations: Civilizations as Superorganisms* (McMann-Fisher Publishing, Auburn, NH, 2010).
______, 2011a, *21st Century Natural Philosophy Of Ultimate Physical Reality* (McMann-Fisher Publishing, Auburn, NH, 2011).
______, 2011b, *All the Universe! Faster Than Light Tachyon Quark Starships & Particle Accelerators with the LHC as a Prototype Starship Drive Scientific Edition* (Pingree-Hill Publishing, Auburn, NH, 2011).
______, 2011c, *From Asynchronous Logic to The Standard Model to Superflight to the Stars* (Blaha Research, Auburn, NH, 2011).

______, 2012a, *From Asynchronous Logic to The Standard Model to Superflight to the Stars volume 2: Superluminal CP and CPT, U(4) Complex General Relativity and The Standard Model, Complex Vierbein General Relativity, Kinetic Theory, Thermodynamics* (Blaha Research, Auburn, NH, 2012).

______, 2012b, *Standard Model Symmetries, And Four And Sixteen Dimension Complex Relativity; The Origin Of Higgs Mass Terms* (Blaha Reasearch, Auburn, NH, 2012).

______, 2013a, *Multi-Stage Space Guns, Micro-Pulse Nuclear Rockets, and Faster-Than-Light Quark-Gluon Ion Drive Starships* (Blaha Research, Auburn, NH, 2013).

______, 2013b, *The Bridge to Dark Matter; A New Sibling Universe; Dark Energy; Inflatons; Quantum Big Bang; Superluminal Physics; An Extended Standard Model Based on Geometry* (Blaha Reasearch, Auburn, NH, 2013).

______, 2014a, *Universes and Megaverses: From a New Standard Model to a Physical Megaverse; The Big Bang; Our Sibling Universe's Wormhole; Origin of the Cosmological Constant, Spatial Asymmetry of the Universe, and its Web of Galaxies; A Baryonic Field between Universes and Particles; Megaverse Extended Wheeler-DeWitt Equation* (Blaha Reasearch, Auburn, NH, 2014).

______, 2014b, *All the Megaverse! Starships Exploring the Endless Universes of the Cosmos Using the Baryonic Force* (Blaha Research, Auburn, NH, 2014).

______, 2014c, *All the Megaverse! II Between Megaverse Universes: Quantum Entanglement Explained by the Megaverse Coherent Baryonic Radiation Devices – PHASERs Neutron Star Megaverse Slingshot Dynamics Spiritual and UFO Events, and the Megaverse Microscopic Entry into the Megaverse* (Blaha Research, Auburn, NH, 2014).

______, 2015a, *PHYSICS IS LOGIC PAINTED ON THE VOID: Origin of Bare Masses and The Standard Model in Logic, U(4) Origin of the Generations, Normal and Dark Baryonic Forces, Dark Matter, Dark Energy, The Big Bang, Complex General Relativity, A Megaverse of Universe Particles* (Blaha Research, Auburn, NH, 2015).

______, 2015b, *PHYSICS IS LOGIC Part II: The Theory of Everything, The Megaverse Theory of Everything, U(4)⊗U(4) Grand Unified Theory (GUT), Inertial Mass = Gravitational Mass, Unified Extended Standard Model and a New Complex General Relativity with Higgs Particles, Generation Group Higgs Particles* (Blaha Research, Auburn, NH, 2015).

____, 2015c, *The Origin of Higgs ("God") Particles and the Higgs Mechanism: Physics is Logic III, Beyond Higgs – A Revamped Theory With a Local Arrow of Time, The Theory of Everything Enhanced, Why Inertial Frames are Special, Universes of the Mind* (Blaha Research, Auburn, NH, 2015).

______, 2015d, *The Origin of the Eight Coupling Constants of The Theory of Everything: U(8) Grand Unified Theory of Everything (GUTE), S^8 Coupling Constant Symmetry, Space-Time Dependent Coupling Constants, Big Bang Vacuum Coupling Constants, Physics is Logic IV* (Blaha Research, Auburn, NH, 2015).

______, 2016a, *New Types of Dark Matter, Big Bang Equipartition, and A New U(4) Symmetry in the Theory of Everything: Equipartition Principle for Fermions, Matter is 83.33% Dark, Penetrating the Veil of the Big Bang, Explicit QFT Quark Confinement and Charmonium, Physics is Logic V* (Blaha Research, Auburn, NH, 2016).

______, 2016b, *The Periodic Table of the 192 Quarks and Leptons in The Theory of Everything: The U(4) Layer Group, Physics is Logic VI* (Blaha Research, Auburn, NH, 2016).

______, 2016c, *New Boson Quantum Field Theory, Dark Matter Dynamics, Dark Matter Fermion Layer Mixing, Genesis of Higgs Particles, New Layer Higgs Masses, Higgs Coupling Constants, Non-Abelian Higgs Gauge Fields, Physics is Logic VII* (Blaha Research, Auburn, NH, 2016).

______, 2016d, *Unification of the Strong Interactions and Gravitation: Quark Confinement Linked to Modified Short-Distance Gravity; Physics is Logic VIII* (Blaha Research, Auburn, NH, 2016).

______, 2016e, *MoND: Unification of the Strong Interactions and Gravitation II, Quark Confinement Linked to Large-Scale Gravity, Physics is Logic IX* (Blaha Research, Auburn, NH, 2016).

______, 2016f, *CQ Mechanics: A Unification of Quantum & Classical Mechanics, Quantum/Semi-Classical Entanglement, Quantum/Classical Path Integrals, Quantum/Classical Chaos* (Blaha Research, Auburn, NH, 2016).

______, 2016g, *GEMS Unified Gravity, ElectroMagnetic and Strong Interactions: Manifest Quark Confinement, A Solution for the Proton Spin Puzzle, Modified Gravity on the Galactic Scale* (Pingree Hill Publishing, Auburn, NH, 2016).

______, 2016h, *Unification of the Seven Boson Interactions based on the Riemann-Christoffel Curvature Tensor* (Pingree Hill Publishing, Auburn, NH, 2016).

______, 2017a, *Unification of the Eleven Boson Interactions based on 'Rotations of Interactions'* (Pingree Hill Publishing, Auburn, NH, 2017).

______, 2017b, *The Origin of Fermions and Bosons, and Their Unification* (Pingree Hill Publishing, Auburn, NH, 2017).

______, 2017c, *Megaverse: The Universe of Universes* (Pingree Hill Publishing, Auburn, NH, 2017).

______, 2017d, *SuperSymmetry and the Unified SuperStandard Model* (Pingree Hill Publishing, Auburn, NH, 2017).

______, 2017e, *From Qubits to the Unified SuperStandard Model with Embedded SuperStrings: A Derivation* (Pingree Hill Publishing, Auburn, NH, 2017).
______, 2017f, *The Unified SuperStandard Model in Our Universe and the Megaverse: Quarks, ...* , (Pingree Hill Publishing, Auburn, NH, 2017).
______, 2018a, *The Unified SuperStandard Model and the Megaverse SECOND EDITION A Deeper Theory based on a New Particle Functional Space that Explicates Quantum Entanglement Spookiness (Volume 1)* (Pingree Hill Publishing, Auburn, NH, 2018).
______, 2018b, *Cosmos Creation: The Unified SuperStandard Model, Volume 2, SECOND EDITION* (Pingree Hill Publishing, Auburn, NH, 2018).
______, 2018c, *God Theory (*Pingree Hill Publishing, Auburn, NH, 2018).
______, 2018d, *Immortal Eye: God Theory: Second Edition* (Pingree Hill Publishing, Auburn, NH, 2018).
______, 2018e, *Unification of God Theory and Unified SuperStandard Model THIRD EDITION* (Pingree Hill Publishing, Auburn, NH, 2018).
______, 2019a, *Calculation of: QED α = 1/137, and Other Coupling Constants of the Unified SuperStandard Theory* (Pingree Hill Publishing, Auburn, NH, 2019).
______, 2019b, *Coupling Constants of the Unified SuperStandard Theory SECOND EDITION* (Pingree Hill Publishing, Auburn, NH, 2019).
______, 2019c, *New Hybrid Quantum Big_Bang–Megaverse_Driven Universe with a Finite Big Bang and an Increasing Hubble Constant* (Pingree Hill Publishing, Auburn, NH, 2019).
______, 2019d, *The Universe, The Electron and The Vacuum* (Pingree Hill Publishing, Auburn, NH, 2019).
______, 2019e, *Quantum Big Bang – Quantum Vacuum Universes (Particles)* (Pingree Hill Publishing, Auburn, NH, 2019).
______, 2019f, *The Exact QED Calculation of the Fine Structure Constant Implies ALL 4D Universes have the Same Physics/Life Prospects* (Pingree Hill Publishing, Auburn, NH, 2019).
______, 2019g, *Unified SuperStandard Theory and the SuperUniverse Model: The Foundation of Science* (Pingree Hill Publishing, Auburn, NH, 2019).
______, 2020a, *Quaternion Unified SuperStandard Theory (The QUeST) and Megaverse Octonion SuperStandard Theory (MOST)* (Pingree Hill Publishing, Auburn, NH, 2020).
______, 2020b, *United Universes Quaternion Universe - Octonion Megaverse* (Pingree Hill Publishing, Auburn, NH, 2020).
______, 2020c, *Unified SuperStandard Theories for Quaternion Universes & The Octonion Megaverse* (Pingree Hill Publishing, Auburn, NH, 2020).
______, 2020d, *The Essence of Eternity: Quaternion & Octonion SuperStandard Theories* (Pingree Hill Publishing, Auburn, NH, 2020).
______, 2020e, *The Essence of Eternity II* (Pingree Hill Publishing, Auburn, NH, 2020).
______, 2020f, *A Very Conscious Universe* (Pingree Hill Publishing, Auburn, NH, 2020).
______, 2020g, *Hypercomplex Universe* (Pingree Hill Publishing, Auburn, NH, 2020).
______, 2020h, *Beneath the Quaternion Universe* (Pingree Hill Publishing, Auburn, NH, 2020).
______, 2020i, *Why is the Universe Real? From Quaternion & Octonion to Real Coordinates* (Pingree Hill Publishing, Auburn, NH, 2020).
______, 2020j, *The Origin of Universes: of Quaternion Unified SuperStandard Theory (QUeST); and of the Octonion Megaverse (UTMOST)* (Pingree Hill Publishing, Auburn, NH, 2020).
______, 2020k, *The Seven Spaces of Creation: Octonion Cosmology* (Pingree Hill Publishing, Auburn, NH, 2020).
______, 2020l, *From Octonion Cosmology to the Unified SuperStandard Theory of Particles* (Pingree Hill Publishing, Auburn, NH, 2020).
______, 2021a, *Pioneering the Cosmos* (Pingree Hill Publishing, Auburn, NH, 2021).
______, 2021b, *Pioneering the Cosmos II* (Pingree Hill Publishing, Auburn, NH, 2021).
______, 2021c, *Beyond Octonion Cosmology* (Pingree Hill Publishing, Auburn, NH, 2021).
______, 2021d, *Universes are Particles* (Pingree Hill Publishing, Auburn, NH, 2021).
______, 2021e, *Octonion-like dna-based life, Universe expansion is decay, Emerging New Physics* (Pingree Hill Publishing, Auburn, NH, 2021).
______, 2021f, *The Science of Creation New Quantum Field Theory of Spaces* (Pingree Hill Publishing, Auburn, NH, 2021).
______, 2021g, *Quantum Space Theory With Application to Octonion Cosmology & Possibly To Fermionic Condensed Matter* (Pingree Hill Publishing, Auburn, NH, 2021).
______, 2021h, *21st Century Natural Philosophy of Octonion Cosmology , and Predestination, Fate, and Free Will* (Pingree Hill Publishing, Auburn, NH, 2021).

______, 2021i, *Beyond Octonion Cosmology II : Origin of the Quantum; A New Generalized Field Theory (GiFT); A Proof of the Spectrum of Universes; Atoms in Higher Universes* (Pingree Hill Publishing, Auburn, NH, 2021).
______, 2021j, *Integration of General Relativity and Quantum Theory: Octonion Cosmology, GiFT, Creation/Annihilation Spaces CASe, Reduction of Spaces to a Few Fermions and Symmetries in Fundamental Frames* (Pingree Hill Publishing, Auburn, NH, 2021).
______, 2022a, *New View of Octonion Cosmology Based on the Unification of General Relativity and Quantum Theory* (Pingree Hill Publishing, Auburn, NH, 2022).
______, 2022b, *The Dust Beneath Hypercomplex Cosmology* (Pingree Hill Publishing, Auburn, NH, 2022).
______, 2022c, *Passing Through Nature to Eternity: ProtoCosmos, HyperCosmos, Unified SuperStandard Theory* (Pingree Hill Publishing, Auburn, NH, 2022).
______, 2022d, *HyperCosmos Fractionation and Fundamental Reference Frame Based Unification: Particle Inner Space Basis of Parton and Dual Resonance Models* (Pingree Hill Publishing, Auburn, NH, 2022).
______, 2022e, *A New UniDimension ProtoCosmos and SuperString F-Theory Relation to the HyperCosmos* (Pingree Hill Publishing, Auburn, NH, 2022).
______, 2022f, *The Cosmic Panorama: ProtoCosmos, HyperCosmos, Unified SuperStandard Theory (UST) Derivation* (Pingree Hill Publishing, Auburn, NH, 2022).
_____, 2022g, *Ultimate Origin: ProtoCosmos and HyperCosmos* (Pingree Hill Publishing, Auburn, NH, 2022).
______, 2023a, *UltraUnification and the Generation of the Cosmos* (Pingree Hill Publishing, Auburn, NH, 2023).
______, 2023b, *God and and Cosmos Theory* (Pingree Hill Publishing, Auburn, NH, 2023).
______, 2023c, *A New Completely Geometric SU(8) Cosmos Theory; New PseudoFermion Fields; Fibonacci-like Dimension Arrays; Ramsey Number Approximation* (Pingree Hill Publishing, Auburn, NH, 2023).
______, 2023d, *Newton's Apple is Now the Fermion* (Pingree Hill Publishing, Auburn, NH, 2023).
______, 2023e,*Cosmos Theory: The Sub-Particle Gambol Model* (Pingree Hill Publishing, Auburn, NH, 2023).
_____, 2024a, *Cosmos-Universe-Particle-Gambol Theory* (Pingree Hill Publishing, Auburn, NH, 2024).
_____, 2024b, *Fractal Cosmos Theory* (Pingree Hill Publishing, Auburn, NH, 2024).
_____, 2024c, *Fractal Cosmic Curve: Tensor-Based CosmosTheory* (Pingree Hill Publishing, Auburn, NH, 2024).
_____, 2024d, *The Eternal Form of Cosmos Theory* (Pingree Hill Publishing, Auburn, NH, 2024).
_____, 2024e, *The Eternal Form of Cosmos Theory Third Edition* (Pingree Hill Publishing, Auburn, NH, 2024).
_____, 2024f, *Fundamental Constants of Cosmos Theory and The Standard Model* (Pingree Hill Publishing, Auburn, NH, 2024).
_____, 2024g, *Quark, Lepton, W and Z Masses of Cosmos Theory and The Standard Model* (Pingree Hill Publishing, Auburn, NH, 2024).
_____, 2024h, *Geometric Cosmos Geometric Universe* (Pingree Hill Publishing, Auburn, NH, 2024).
_____, 2024i, *Particles and Universes of Cosmos Theory* (Pingree Hill Publishing, Auburn, NH, 2024).
_____, 2024j, *Unification of the Subluminal and the Superluminal in Cosmos Theory* (Pingree Hill Publishing, Auburn, NH, 2024).
_____, 2024k, *The Dawn of Dynamic Cosmos Dimension Arrays* (Pingree Hill Publishing, Auburn, NH, 2024).
_____, 2024ℓ, *Structure and Dynamics of Cosmos Theory and the Unified SuperStandard Theory* (Pingree Hill Publishing, Auburn, NH, 2024).
_____, 2025a, *Black Holes, White Holes, and Superluminal Starship* (Pingree Hill Publishing, Auburn, NH, 2025).
_____, 2025b, *The Sequence Theory Foundation of Cosmos Theory and Gambol Theory* (Pingree Hill Publishing, Auburn, NH, 2025).
____, 2025c, *BEYOND TIME AND SPACE Fermion Strata & Consistency Conditions for Sequences, Cosmos Spaces and Gambols* (Pingree Hill Publishing, Auburn, NH, 2025).
____, 2025d, *The C o s m o s as a Physics Organism* (Blaha Research, Auburn, NH, 2025).
____, 2025e, *The Root of Reality* (Blaha Research, Auburn, NH, 2025).
Eddington, A. S., 1952, *The Mathematical Theory of Relativity* (Cambridge University Press, Cambridge, U.K., 1952).
Fant, Karl M., 2005, *Logically Determined Design: Clockless System Design With NULL Convention Logic* (John Wiley and Sons, Hoboken, NJ, 2005).
Feinberg, G. and Shapiro, R., 1980, *Life Beyond Earth: The Intelligent Earthlings Guide to Life in the Universe* (William Morrow and Company, New York, 1980).
Gelfand, I. M., Fomin, S. V., Silverman, R. A. (tr), 2000, *Calculus of Variations* (Dover Publications, Mineola, NY, 2000).
Giaquinta, M., Modica, G., Souchek, J., 1998, *Cartesian Coordinates in the Calculus of Variations* Volumes I and II (Springer-Verlag, New York, 1998).
Giaquinta, M., Hildebrandt, S., 1996, *Calculus of Variations* Volumes I and II (Springer-Verlag, New York, 1996).

Gradshteyn, I. S. and Ryzhik, I. M., 1965, *Table of Integrals, Series, and Products* (Academic Press, New York, 1965).
Heitler, W., 1954, *The Quantum Theory of Radiation* (Claendon Press, Oxford, UK, 1954).
Huang, Kerson, 1992, *Quarks, Leptons & Gauge Fields 2nd Edition* (World Scientific Publishing Company, Singapore, 1992).
Jost, J., Li-Jost, X., 1998, *Calculus of Variations* (Cambridge University Press, New York, 1998).
Kaku, Michio, 1993, *Quantum Field Theory*, (Oxford University Press, New York, 1993).
Kirk, G. S. and Raven, J. E., 1962, *The Presocratic Philosophers* (Cambridge University Press, New York, 1962)
Landau, L. D. and Lifshitz, E. M., 1987, *Fluid Mechanics 2nd Edition*, (Pergamon Press, Elmsford, NY, 1987).
Rescher, N., 1967, *The Philosophy of Leibniz* (Prentice-Hall, Englewood Cliffs, NJ, 1967).
Riesz, Frigyes and Sz.-Nagy, Béla, 1990, *Functional Analysis* (Dover Publications, New York, 1990).
Sakurai, J. J., 1964, *Invariance Principles and Elementary Particles* (Princeton University Press, Princeton, NJ, 1964).
Weinberg, S., 1972, *Gravitation and Cosmology* (John Wiley and Sons, New York, 1972).
Weinberg, S., 1995, *The Quantum Theory of Fields Volume I* (Cambridge University Press, New York, 1995).

INDEX

About the Author

Stephen Blaha is a well-known Physicist and Man of Letters with interests in Science, Society and civilization, the Arts, and Technology. He had an Alfred P. Sloan Foundation scholarship in college. He received his Ph.D. in Physics from Rockefeller University. He has served on the faculties of several major universities. He was also a Member of the Technical Staff at Bell Laboratories, a manager at the Boston Globe Newspaper, a Director at Wang Laboratories, and President of Blaha Software Inc. and of Janus Associates Inc. (NH).

Among other achievements he was a co-discoverer of the "r potential" for heavy quark binding developing the first (and still the only demonstrable) non-Aeolian gauge theory with an "r" potential; first suggested the existence of topological structures in superfluid He-3; first proposed Yang-Mills theories would appear in condensed matter phenomena with non-scalar order parameters; first developed a grammar-based formalism for quantum computers and applied it to elementary particle theories; first developed a new form of quantum field theory without divergences (thus solving a major 60 year old problem that enabled a unified theory of the Standard Model and Quantum Gravity without divergences to be developed); first developed a formulation of complex General Relativity based on analytic continuation from real space-time; first developed a generalized non-homogeneous Robertson-Walker metric that enabled a quantum theory of the Big Bang to be developed without singularities at t = 0; first generalized Cauchy's theorem and Gauss' theorem to complex, curved multi-dimensional spaces; received Honorable Mention in the Gravity Research Foundation Essay Competition in 1978; first developed a physically acceptable theory of faster-than-light particles; first derived a composition of extremums method in the Calculus of Variations; first quantitatively suggested that inflationary periods in the history of the universe were not needed; first proved Gödel's Theorem implies Nature must be quantum; provided a new alternative to the Higgs Mechanism, and Higgs particles, to generate masses; first showed how to resolve logical paradoxes including Gödel's Undecidability Theorem by developing Operator Logic and Quantum Operator Logic; first developed a quantitative harmonic oscillator-like model of the life cycle, and interactions, of civilizations; first showed how equations describing superorganisms also apply to civilizations. A recent book shows his theory applies successfully to the past 14 years of history and to *new* archaeological data on Andean and Mayan civilizations as well as Early Anatolian and Egyptian civilizations.

He first developed an axiomatic derivation of the form of The Standard Model from geometry – space-time properties – The Unified SuperStandard Model. It unifies all the known forces of Nature. It also has a Dark Matter sector that includes a Dark ElectroWeak sector with Dark doublets and Dark gauge interactions. It uses quantum coordinates to remove infinities that crop up in most interacting quantum field theories and additionally to remove the infinities that appear in the Big Bang and generate inflationary growth of the universe. It shows gravity has a MOND-like form without sacrificing Newton's Laws. It relates the interactions of the MOND-like sector of gravity with the r-potential of Quark Confinement. The axioms of the theory lead to the question of their origin. We suggest in the preceding edition of this book it can be attributed to an entity with God-like properties. We explore these properties in "God Theory" and show they predict that the Cosmos exists forever although individual universes (or incarnations of our universe) "come and go." Several other important results emerge from God Theory such a functionally triune God. The Unified SuperStandard Theory has many other important parts described in the Current Edition of *The Unified SuperStandard Theory* and expanded in subsequent volumes.

Blaha has had a major impact on a succession of elementary particle theories: his Ph.D. thesis (1970), and papers, showed that quantum field theory calculations to all orders in ladder approximations could not give scaling deep inelastic electron-nucleon scattering. He later showed the eigenvalue equation for the fine structure constant α in Johnson-Baker-Willey QED had a zero at α = 1 not 1/137 by solving the Schwinger-Dyson equations to all orders in an approximation that agreed with exact results to 4th order in α thus ending interest in this theory. In 1979 at Prof. Ken Johnson's (MIT) suggestion he calculated the proton-neutron mass difference in the MIT bag model and found the result had the wrong sign reducing interest in the bag model. These results all appear in Physical Review papers. In the 2000's he repeatedly pointed out the shortcomings of SuperString theory and showed that The Standard Model's form could be derived from space-time geometry by an extension of Lorentz transformations to faster than light transformations. This deeper space-time basis greatly increases the possibility that it is part of THE fundamental theory. Recently, Blaha showed that the Weak interactions differed significantly from the Strong, electromagnetic and gravitation interactions in important respects while these interactions had similar features, and suggested that ElectroWeak theory, which is essentially a glued union of the Weak interactions and Electromagnetism, possibly modulo unknown Higgs particle features, be replaced by a unified theory of the other interactions combined with a stand-alone Weak interaction theory. Blaha also showed that, if Charmonium calculations are taken seriously, the Strong interaction coupling constant is only a factor of five larger than the electromagnetic coupling constant, and thus Strong interaction perturbation theory would make sense and yield physically meaningful results.

In graduate school (1965-71) he wrote substantial papers in elementary particles and group theory: The Inelastic E- P Structure Functions in a Gluon Model. Phys. Lett. B40:501-502,1972; Deep-Inelastic E-P Structure Functions In A Ladder Model With Spin 1/2 Nucleons, Phys.Rev. D3:510-523,1971; Continuum Contributions To The Pion Radius, Phys. Rev. 178:2167-2169,1969; Character Analysis of U(N) and SU(N), J. Math. Phys. 10, 2156 (1969); and The Calculation of the Irreducible Characters of the Symmetric Group in Terms of the Compound Characters, (Published as Blaha's Lemma in D. E. Knuth's book: *The Art of Computer Programming Vols. 1 – 4*).

In the early 1980's Blaha was also a pioneer in the development of UNIX for financial, scientific and Internet applications: benchmarked UNIX versions showing that block size was critical for UNIX performance, developing financial modeling software, starting database benchmarking comparison studies, developing Internet-like UNIX networking (1982) and developing a hybrid shell programming technique (1982) that was a precursor to the PERL programming language. He was also the manager of the AT&T ten-year future products development database. His work helped lead to commercial UNIX on computers such as Sun Micros, IBM AIX minis, and Apple computers.

In the 1980's he pioneered the development of PC Desktop Publishing on laser printers and was nominated for three "Awards for Technical Excellence" in 1987 by PC Magazine for PC software products that he designed and developed.

Recently he has developed a theory of Megaverses – actual universes of which our universe is one – with quantum particle-like properties based on the Wheeler-DeWitt equation of Quantum Gravity. He has developed a theory of a baryonic force, which had been conjectured many years ago, and estimated the strength of the force based on discrepancies in measurements of the gravitational constant G. This force, operative in D-dimensional space, can be used to escape from our universe in "uniships" which are the equivalent of the faster-than-light starships proposed in the author's earlier books. Thus travel to other universes, as well as to other stars is possible.

Blaha also considered the complexified Wheeler-DeWitt equation and showed that its limitation to real-valued coordinates and metrics generated a Cosmological Constant in the Einstein equations.

The author has also recently written a series of books on the serious problems of the United States and their solution as well as a book on the decline of Mankind that will follow from current social and genetic trends in Mankind.

In the past twenty years Dr. Blaha has written over 80 books on a wide range of topics. Some recent major works are: *From Asynchronous Logic to The Standard Model to Superflight to the Stars, All the Universe!, SuperCivilizations: Civilizations as Superorganisms, America's Future: an Islamic Surge, ISIS, al Qaeda, World Epidemics, Ukraine, Russia-China Pact, US Leadership Crisis, The Rises and Falls of Man – Destiny – 3000 AD: New Support for a Superorganism MACRO-THEORY of CIVILIZATIONS From CURRENT WORLD TRENDS and NEW Peruvian, Pre-Mayan, Mayan, Anatolian, and Early Egyptian Data, with a Projection to 3000 AD,* and *Mankind in Decline: Genetic Disasters, Human-Animal Hybrids, Overpopulation, Pollution, Global Warming, Food and Water Shortages, Desertification, Poverty, Rising Violence, Genocide, Epidemics, Wars, Leadership Failure.*

He has taught approximately 4,000 students in undergraduate, graduate, and postgraduate corporate education courses primarily in major universities, and large companies and government agencies.

He developed a quantum theory, The Unified SuperStandard Theory (UST), which describes elementary particles in detail without the difficulties of conventional quantum field theory. He found that the internal symmetries of this theory could be exactly derived from an octonion theory called QUeST. He further found that another octonion theory (UTMOST) describes the Megaverse. It can hold QUeST universes such as our own universe. It has an internal symmetry structure which is a superset of the QUeST internal symmetries.

Recently he developed Octonion Cosmology. He replaced it with HyperCosmos theory, which has significantly better features. He developed a fractionalization process for dimensions, particles and symmetry groups. He also described transformation that reduced particlcs and dimensions to a far more compact form. He also developed a precursor theory ProtoCosmos that leads to the HyperCosmos.

The author showed that space-time and Internal Symmetries can be unified in any of the ten HyperCosmos spaces in their associated HyperUnification spaces. The combined set of HyperUnification spaces enable all HyperCosmos dimensions to be obtained by a General Relativistic transformation from one primordial dimension in the 42 space-time dimension unified HyperUnification space.

At present the author devel;oped the Cosmos Theory that incorporates ProtoCosmos Theory, HyperCosmos Theory, Limos Theory, Second Kind HyperCosmos Theory and HyperUnification Spaces. He has introduced PseudoFermion wave functions and theory, He has related Cosmos Theory to Regge trajectories of spaces, parton theory, Veneziano amplitudes, Fibonacci numbers and Ramsey numbers. He has calculated an approximation to the difficult R(n,n) Ramsey numbers.

He has developed a Gambol Model that successfully accounts for e-p deep inelastic scattering, fundamental particle resonances, hadron scattcring, and the inner structure of particles based on confinement through Casimir forces of ideal gambol gases. The Gambol Planckian Distribution was derived.

He has applied the Gambol Model to particles, universes, and the Cosmos of universes. He showed that the Cosmos may have a distribution of 23 universes corresponding to various Cosmos spaces.

Recently he showed that Cosmos Theory follows from the number of independent asymmetric tensors in a dimension r. He also showed the close parallel between the form of γ-matrices and Cosmos Theory dimension arrays. The closeness suggested that dimension arrays have the same importance as γ-matrices for fermions.

He demonstrated that the pressure of fermions within a space of dimension r balances the Casimir vacuum energy force for 18 dimensions. He showed that $2e\pi = 17.02$ marks the critical point where pressure balances Casimir force, which implies $r = 18$ is the highest dimension Physical Cosmos space. The dimension $2e\pi$ appears to set the approximate dimension for Cosmos spaces with dimension array size $2^{r+4} \cong (17.02/8)^{r+4}. \cong (e\pi/4)^{r+4}. \cong 2.13^{r+4}$.

Now he has found the sequences of Coupling Constant values and fermion masses in the Standard Model and UST. This book unifies the Cosmos spaces spectrum, the coupling constant spectrum and the fermion mass sequences in a fundamental scaling differential equation.

This book shows the fractal basis of Cosmos Theory.

www.ingramcontent.com/pod-product-compliance
Lightning Source LLC
LaVergne TN
LVHW071339210926
848486LV00003B/192
* 9 7 9 8 9 9 8 6 0 7 1 4 1 *